Building Cross-Platform GUI Applications with Fyne

Create beautiful, platform-agnostic graphical applications using Fyne and the Go programming language

Andrew Williams

BIRMINGHAM—MUMBAI

Building Cross-Platform GUI Applications with Fyne

Group Product Manager: Aaron Lazar
Publishing Product Manager: Richa Tripathi
Senior Editor: Nitee Shetty
Content Development Editor: Tiksha Lad
Technical Editor: Gaurav Gala
Copy Editor: Safis Editing
Project Coordinator: Deeksha Thakkar
Proofreader: Safis Editing
Indexer: Tejal Daruwale Soni
Production Designer: Joshua Misquitta

First published: January 2021
Production reference: 2300321

Published by Packt Publishing Ltd.
Livery Place
35 Livery Street
Birmingham
B3 2PB, UK.

ISBN 978-1-80056-316-2

www.packt.com

To the fantastic Fyne community, everyone involved in the project, or who uses it in their own work. Without your contributions and involvement, there would be no toolkit to write about.

– Andrew Williams

Contributors

About the author

Andrew Williams graduated from the University of Edinburgh in 2003 with a bachelor's degree, with honors, in computer science. After university, he went to work as a software engineer and has gained over 15 years of commercial software development experience across a variety of programming languages, including Java, C, Objective-C, and Go. Andrew has spent many years working as a CTO with many early-stage and growing software start-ups. He has been a core developer in large open source projects, including Enlightenment, EFL, and Maven, as well as involved in maintaining various community websites and tutorials. Andrew's passion for building tools and services that make software development simpler led him to start authoring books on the subject.

Endless thanks go to my wife, Michelle, for supporting and encouraging me through the development of Fyne and of this book. Many thanks also go to my friends and family for providing motivational messages throughout a challenging year. Thanks to my excellent reviewer, Jacob Alzén, for his time and dedication in making this book the best it could be.

About the reviewer

Jacob Alzén is an engineering student, programmer, and privacy advocate from Sweden. He started out his software career as a volunteer doing community support and release testing for Brave Software and later translated the whole browser into Swedish. This sparked an interest in privacy issues that resulted in the primary usage of open source software on Linux. He has more than five years of Linux experience along with a couple of years of experience in C++ and Go programming. Jacob has also taught C++ programming at both beginner and intermediate level during his own studies at the Curt Nicolin Gymnasium in Sweden.

Table of Contents

Section 2: Components of a Fyne App

3

Window, Canvas, and Drawing

4

Layout and File Handling

5

Widget Library and Themes

6

Data Binding and Storage

7

Building Custom Widgets and Themes

Section 3: Packaging and Distribution

8

Project Structure and Best Practices

9

Bundling Resources and Preparing for Release

10

Distribution – App Stores and Beyond

Appendix A: Developer Tool Installation

Appendix B: Installing Mobile Build Tools

Appendix C: Cross-Compiling

Other Books You May Enjoy

Index

Preface

The development of graphical user interfaces has a long history, resulting in a complicated landscape today. There are many different ways to build applications, each with their own benefits and drawbacks. In this modern world, with so many different devices in our everyday lives, it can seem impossible to reach an entire audience without building many different apps or GUIs. The technologies required are often platform-specific, and many of the technologies that exist have evolved across decades, leaving a legacy that can slow down new developers and experienced teams alike.

In the same way that the Go programming language was designed to make software development easier across all operating systems, the Fyne toolkit aims to empower graphical app creation in a platform-agnostic manner. This definitive guide to building applications with the Fyne toolkit aims to assist software engineers of any experience level in learning the APIs and processes involved. From your first line of Fyne code through to deploying an application to global marketplaces, the samples and screenshots will guide you on each step of the journey.

Who this book is for

This book is written both for Go developers who are interested in building native graphical applications and for those with platform-specific GUI experience looking for a cross-platform solution. Some knowledge of building Go applications is assumed, but not essential. The book opens with a history of the GUI for anyone who is not familiar, and then introduces the Fyne project and its vision and ambition to solve many of the challenges faced by developers of native applications across desktop and mobile devices.

With the help of code snippets and worked examples, developers of every level should be successful in building their first Fyne apps. As well as running these apps on your computers and mobile devices, you will also be taken through the preparation and upload process to deploy to app stores and marketplaces.

What this book covers

Chapter 1, A Brief History of GUI Toolkits and Cross-Platform Development, contains a short reminder of the history behind graphical applications and how the toolkits used to develop them have evolved over time. We look at the different approaches to cross-platform development and why it is important. By the end of this chapter, you will be familiar with the benefits and challenges of GUI toolkits in approaching cross-platform app development.

Chapter 2, The Future According to Fyne, introduces the Fyne toolkit and its approach to supporting all operating systems, with a Material Design-inspired user interface look and feel. In this chapter, we will explore the vision of the Fyne toolkit and how it builds on the Go language to create an easy-to-use, cross-platform GUI toolkit. After reading this chapter, you will see how Fyne aims to solve the challenges outlined in the first chapter of the book and how it aims to shape app development in the future.

Chapter 3, Window, Canvas, and Drawing, introduces the main APIs behind the rendering layer of the Fyne toolkit. We will see how to draw objects and how to combine them using containers to create more complex output. This chapter also covers the APIs required to load an application and manage its windows. We complete this chapter with an example that uses graphic elements and animations to create a simple cross-platform game.

Chapter 4, Layout and File Handling, expands on the manual placement of elements in the previous chapter. We will examine the standard layouts that are provided and how they are combined to form complex user interface structures, as well as how to build our own. Also covered is the filesystem abstraction, which provides standard file access for traditional filesystems and the more complex mobile data sharing methodologies. We will apply all this knowledge to create an image browsing app at the end of the chapter.

Chapter 5, Widget Library and Themes, introduces the largest package within the Fyne toolkit – its widget library. In this chapter, we will explore the main widgets available and how to use them in constructing an application GUI. We will see how their standardized look and feel can be influenced through theme selection and how user preferences, such as light or dark modes, are rendered using the built-in theme. This chapter is completed with a step-by-step creation of a task management app using many of the widgets explored earlier.

Chapter 6, Data Binding and Storage, explores APIs that help to efficiently manage data and storage within a Fyne app. We will see how the widgets seen in the previous chapter can be bound to data elements and thereby avoid much of the code that was necessary to set up and manage their contents. Also demonstrated is how applications can manage user preferences and how they can connect through data binding directly to widget values. The concepts in this chapter are applied through the creation of a health app that helps to track your water consumption.

Chapter 7, Building Custom Widgets and Themes, demonstrates how applications with bespoke requirements can build on the solid foundations explored in the previous chapters. We will examine the various ways in which developers can customize and extend existing widgets, or build completely bespoke components. We will also see how custom themes can be loaded to give apps more brand identity or add custom fonts and icons. Using these features, we will build an instant messenger user interface that displays a distinct style and custom widgets.

Chapter 8, Project Structure and Best Practices, builds on the best practices that are well documented for the Go language. We will see how to organize projects to keep the code clean and facilitate easier maintenance as they grow. We will also explore how unit testing and test-driven development are possible and encouraged when building GUIs with Fyne. Additionally, on the rare occasion where platform-specific code may be required, we see how an app can adjust its behavior for different target platforms.

Chapter 9, Bundling Resources and Preparing for Release, explains how packaging up a graphical application is more complex than a simple binary that would be accessed from the command line. In this chapter, we see what metadata is required to allow a Fyne app to blend in with other desktop and mobile applications. We also step through the process for bundling apps so that they can be shared and installed, as end users would expect of a platform native application.

Chapter 10, Distribution – App Stores and Beyond, confronts the final challenge of cross-platform application development – distribution. We will see how applications are prepared for public release and packaged with extra data and code signing required by many app stores. We wrap up the book by stepping through the upload process for Apple, Google, and Microsoft stores, along with distribution to Unix systems.

Appendix A, Developer Tool Installation, contains platform-specific steps for managing software required by other chapters. This chapter will assist new developers to install the compilers and tools required to build the examples in this book.

Appendix B, Installation of Mobile Build Tools, provides documentation for setting up the additional tools that will be required to build Fyne apps for mobile devices. Android and iOS builds can be created locally after following these steps. Alternatively developers could choose to use fyne-cross, covered in Appendix C, Cross-Compiling.

Appendix C, Cross-Compiling, outlines the platform specific installation and configuration that is required to set up the building of different operating systems. Following these steps, developers will be able to compile their Fyne apps for different platforms using just one computer. It also shows how to set up the Fyne cross compiling solution fyne-cross for developers that don't want to manage the tools in detail themselves.

To get the most out of this book

A basic knowledge of the Go language is assumed throughout this book. If you are not yet familiar with the syntax or concepts, consider running through the online tutorial before you begin reading (tour.golang.org).

To run the examples, you will need at least version 1.12 of Go installed, as well as a C compiler for your computer (required by the Fyne library code). If either of these are not installed, you can find detailed steps in *Appendix A, Developer Tool Installation*. We will step through the process of installing the Fyne library and its supporting tools as they are required throughout the book:

Software covered in the book	OS requirements
Go 1.12 (or newer)	Windows, macOS, or Linux computer for development
Fyne 2.0	iOS, iPadOS, or Android device for mobile testing (optional)

To develop applications, you will need a Windows, macOS, or Linux computer with the development tools described previously. To test applications on mobile devices, you will also need the Android SDK and/or Xcode (for iOS/iPadOS) environment installed. For more information, refer to *Appendix B, Installation of Mobile Build Tools*. For the further testing of mobile builds, it is advisable to have a suitable mobile device available.

To derive additional benefits from this book, it would be ideal if you have in mind an app project you are aiming to build. Doing so will help to practice the example code in different settings to better understand the widget and API capabilities. Additionally, this will mean that the app store upload process at the end of this book will result in a published version of your application!

If you are using the digital version of this book, we advise you to access the code via the GitHub repository (link available in the next section). Doing so will help you avoid any potential errors related to the copying and pasting of code.

Download the example code files

You can download the example code files for this book from GitHub at https://github.com/PacktPublishing/Building-Cross-Platform-GUI-Applications-with-Fyne. In case there's an update to the code, it will be updated on the existing GitHub repository.

We also have other code bundles from our rich catalog of books and videos available at https://github.com/PacktPublishing/. Check them out!

Download the color images

We also provide a PDF file that has color images of the screenshots/diagrams used in this book. You can download it here: https://static.packt-cdn.com/downloads/9781800563162_ColorImages.pdf

Conventions used

There are a number of text conventions used throughout this book.

Code in text: Indicates code words in text, database table names, folder names, filenames, file extensions, pathnames, dummy URLs, user input, and Twitter handles. Here is an example: "Using method-based updates, SetText() and SetIcon() calls will refresh the widget, possibly triggering the preceding issue."

A block of code is set as follows:

```
const (
    serverKeyDevelopment = "DEVELOPMENT_KEY"
    serverKeyProduction  = "PRODUCTION_KEY"
)
```

Any command-line input or output is written as follows:

```
$ fyne release -appVersion 1.0 -appBuild 1 -certificate
"CertificateName" -profile "ProfileName"
```

When a command relates to files or data within the GitHub repository (linked above and at the top of each chapter), the command prompt will start with the directory name, as follows:

```
Chapter03/window$ go run main.go
```

Bold: Indicates a new term, an important word, or words that you see on screen. For example, words in menus or dialog boxes appear in the text like this. Here is an example: "You can also find the **Settings** panel in the **Settings** menu within **fyne_demo**."

> **Tips or important notes**
> Appear like this.

Get in touch

Feedback from our readers is always welcome.

General feedback: If you have questions about any aspect of this book, mention the book title in the subject of your message and email us at customercare@packtpub.com.

Errata: Although we have taken every care to ensure the accuracy of our content, mistakes do happen. If you have found a mistake in this book, we would be grateful if you would report this to us. Please visit www.packtpub.com/support/errata, selecting your book, clicking on the Errata Submission Form link, and entering the details.

Piracy: If you come across any illegal copies of our works in any form on the internet, we would be grateful if you would provide us with the location address or website name. Please contact us at copyright@packt.com with a link to the material.

If you are interested in becoming an author: If there is a topic that you have expertise in, and you are interested in either writing or contributing to a book, please visit authors.packtpub.com.

Reviews

Please leave a review. Once you have read and used this book, why not leave a review on the site that you purchased it from? Potential readers can then see and use your unbiased opinion to make purchase decisions, we at Packt can understand what you think about our products, and our authors can see your feedback on their book. Thank you!

For more information about Packt, please visit packt.com.

Section 1: Why Fyne? The Reason for Being and a Vision of the Future

Since their invention nearly 50 years ago, **graphical user interfaces** (**GUIs**) have been the standard way to interact with a software product. In this time they have evolved, and traditional graphical applications are being challenged by the ubiquity of web-based software and new interaction methods for smartphones and handheld computers. Despite these new trends, there are still many reasons why building a native graphical application could be the right strategy for your product—especially if it could be deployed to all the platforms available.

In this section, we will see how GUIs and the graphical toolkits used to program them have evolved. We will explore the pros and cons of the technologies over the years and how some can be used for cross-platform development. We will learn about the Fyne project, its background and vision, and how it aims to be an ideal solution to the evolving needs of GUI development.

This section will cover the following topics:

- *Chapter 1, A Brief History of GUI Toolkits and Cross-Platform Development*
- *Chapter 2, The Future According to Fyne*

We start with a short review of the history of desktop computers and traditional GUIs.

1

A Brief History of GUI Toolkits and Cross-Platform Development

This book is aimed at exploring how to easily build robust and beautiful graphical applications that will work well across all operating systems and devices. Before we start looking at the details of how this is accomplished, it is important to consider the history of these devices and the landscape of graphical toolkits throughout the last 50 years. We start with a reminder of where GUI-based applications started and how far they have come.

In this chapter, you will be reintroduced to the **Graphical User Interface** (**GUI**), along with learning about toolkits that support app development and how they offer different approaches to cross-platform development. We will explore the benefits of coding a native GUI for responsive user experience and platform integration. Upon completion of this chapter, you should be familiar with the origins and challenges of graphical toolkits and the different approaches that have been taken during this journey.

In this chapter, we'll cover the following topics to provide a short history of GUI toolkits and cross-platform development:

- Where GUI toolkits came from
- How they have adapted (or stayed the same) over time
- Historical approaches to cross-platform development

Understanding the history of the graphical user interface

In 1973, the *Palo Alto Research Center* (*Xerox PARC*) completed the Alto computer, the first commercial example of a graphical desktop computer. In most contemporary histories, this was the first example of what we understand as the GUI. While the screen orientation and lack of colors make it a little peculiar to the modern eye, it's clearly recognizable and includes many key components as well as a mouse and keyboard for interaction. While it took another 7 years to become generally available to the public in 1981 as the *Xerox Star*, it was clear that this was a dramatic step forward:

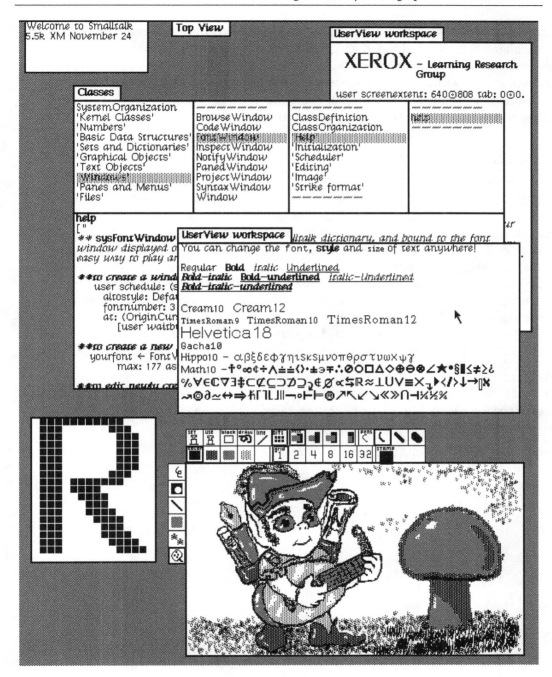

Figure 1.1 – Dynabook environment desktop (1976; Smalltalk-76 running on Alto). Copyright SUMIM.
ST, licensed CC BY-SA 4.0

This development was a huge leap forward for the usability of computers. Up to this time, all interaction was through text-mode computer screens and keyboard or other text input devices. The graphical interface is much easier to learn for a novice looking to get started, and allows the quicker discovery of advanced features. While the command-line interface remains popular with programmers and other *pro users*, the GUI is the largest contributing factor to the rise of the desktop computer.

Popularity of the desktop computer

The introduction of a user-friendly graphical environment brought about a significant growth in the use of desktop computers. Around the time of the Alto computer, there were an estimated 48,000 desktop computers around the world. By 2001, this number had increased dramatically to over 125,000,000 personal computers shipped (https://en.wikipedia.org/wiki/History_of_personal_computers#Market_size). In 2002, the industry celebrated a billion computers shipped (http://news.bbc.co.uk/1/hi/sci/tech/2077986.stm), though numbers have declined more recently (see the *Smartphones and mobile apps* section later in this chapter) and fewer than 300 million were reported shipped in 2018 (https://venturebeat.com/2019/01/10/gartner-and-idc-hp-and-lenovo-shipped-the-most-pcs-in-2018-but-total-numbers-fell/).

As these devices reached the hands of consumers, and hardware became more capable, we started to see a focus on creating attractive user interfaces as well as a trend to establish or match fashion trends. The following are some important versions of Microsoft's Windows operating system:

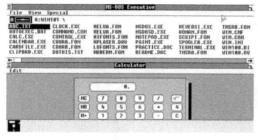

Figure 1.2 – Windows 1 (1985)

Figure 1.3 – Windows 3.11 (1993)

Figure 1.4 – Windows XP (2001)

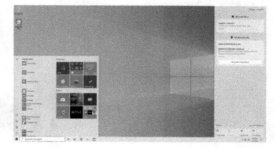

Figure 1.5 – Windows 10 (2015)

The copyright for all screenshots belongs to Microsoft. Each image used with permission

As you can see in the previous screenshots, each major revision of the desktop environment brought new styles for the buttons, fonts, and other user interface elements. This is all controlled by the toolkit and represents an evolution in usability and style choices that we'll explore later in this chapter.

Whilst Microsoft were progressing with their GUIs, there were also many competitors, some of which may appear familiar and others with their own distinct styles; for example, the following popular systems:

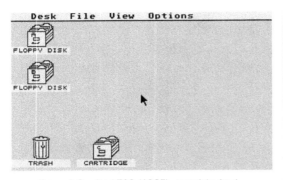

Figure 1.6 – Atari TOS (1985), copyright Atari

Figure 1.7 – BeOS (1995), copyright Palm

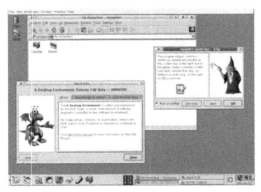

Figure 1.8 – KDE 2 (2000), copyright KDE community

Figure 1.9 – MacOS El Capitan (2015), copyright Apple

Desktop screenshots from various operating systems 1985-2015. Each image has been used with the required permission under fair use policies

As you can see from the preceding desktop screenshots for various operating systems from 1985 to 2015, there have been dramatic shifts in the look and feel, whilst maintaining a certain familiarity. These desktop systems are all designed for running multiple application windows typically centered around document editing, file management, and utility apps. Additional software, such as games, photo management, and music players, appeared over the years, but the most ubiquitous, the web browser, was not commonplace until the late 1990s. The addition of internet access started a shift to a new era of computing.

Moving to the web

With the increasing availability of reliable internet connections, we started to see an increase in the amount of information being accessed from servers on the **World Wide Web (WWW)**. Providing a good user web browser experience became of paramount importance and the fierce competition saw operating system manufacturers up against independent software developers (search `browser wars` in your favorite search engine to know more).

The WWW was first proposed by Sir Tim Berners Lee in 1981 and development began within the CERN (`https://home.cern`) project (codenamed *ENQUIRE*). The early web was made available to the public in 1993. As a distributed system that anyone is able to add to, innovation in design was even more rapid than in the desktop operating systems we saw earlier. Trends in design and usability quickly caught up with, overtook, and started to lead traditional software development:

Figure 1.10 – The first website (1991), Tim Berners Lee

Figure 1.11 – GeoCities home page (1994)

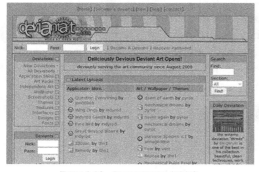

Figure 1.12 – DeviantArt site (2000)

Figure 1.13 – Bootstrap site design framework (2011)

Trends in website design (via Web Design Museum); copyrights belong to their respective owners

The web started as a project for providing access to data, born out of a frustration of how difficult it could be to access information on a different computer. What started as simple information retrieval quickly became a polished presentation of more complex information, and then began to become a place to submit or manipulate information as well.

A simple data access platform quickly grew into much more and before long, this emerged as a full application platform. In fact, due to the standards-based approach (overseen by the **World Wide Web Consortium (W3C)**), this was one of the first truly cross-platform development opportunities. A web-based application would be developed and made available to all computers in one go—a big advancement over previous attempts to develop for multiple platforms.

An additional benefit of delivering applications through a web-based solution was that you could support multiple types of application accessing the underlying data or functionality. A web-based **API (Application Programming Interface)** that had historically powered the user-visible website could be used by other devices as well. This design allowed traditional software to access the same data as the web-based delivery systems and contributed to the development of common-place architectures that support a multitude of different types of software – including the more recent mobile-based applications.

Smartphones and mobile apps

In 2007, Apple's Steve Jobs introduced the iPhone, a fresh new design for the concept of mobile computing. Although portable *smartphone* devices existed for many years before this event, the introduction of a slick new user interface, touchscreen input, and large screen for displaying video and web content had a significant impact on the market. Competitors (existing and newly created) were now racing to create the best user experience that could fit in the consumers' pockets. Although early devices touted that you could browse any website with ease, developers quickly adapted the content to be better presented on these smaller screens—often focusing on information that mattered on the move.

To satisfy the user demand for a more sophisticated and faster experience on these more limited (by hardware or internet connectivity) devices, the concept of a *mobile app* was born. These small pieces of software were designed specifically for a certain type of mobile phone (Android, iPhone, and others) and were made available by the platform's store or marketplace. Such software had large benefits over the web-based solutions that came earlier as they could be installed on the device, so they ran faster and were developed specifically for the given hardware, creating a better user experience and allowing access to the more advanced capabilities of each device (such as location detection, thumbprint sensors, and Bluetooth).

These native apps provided the ultimate user experience. The applications could be very fast (as they were installed on the device), adapt to the user (through access to local settings and data), and also interact with the operating system features (such as calendars, voice controls, and cutting-edge hardware sensors), all of which are not really possible when delivered through a web app. However, they came with a disadvantage for developers—not only did each platform look different, meaning that designs may need to be adapted, but they were also distributed separately and typically required different programming languages to develop. Now instead of reaching the whole world with a single application, a software company would need at least three different apps to reach their customers through their favorite devices:

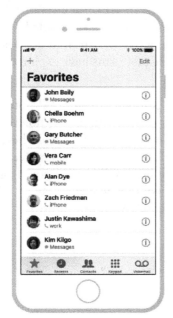 

Figure 1.14 – An iPhone app, copyright Apple   Figure 1.15 – An Android app, copyright Google

iPhone and Android devices showing their similarities and differences

We will come back to the challenges of developing for multiple different hardware platforms later, but first, we will explore the graphical toolkits that underpinned the various technologies we have seen in this section.

Exploring the evolution of GUI toolkits

GUIs must be programmed like any other computer program, and just like libraries are created to provide standard components, a GUI toolkit exists to support building the graphical elements of an application. Many toolkits exist for different reasons—Wikipedia maintains a list of nearly 50 different projects, and growing, at `https://en.wikipedia.org/wiki/List_of_widget_toolkits` and `https://en.wikipedia.org/wiki/List_of_platform-independent_GUI_libraries`. To make sense of the huge number of options, we split them into categories, looking first at those built for specific operating systems.

Platform-specific toolkits

Each graphical operating system or desktop environment has a distinct look and programming style, therefore a graphical toolkit was traditionally created for each platform. Windows had the WinAPI (as well as WinForms and foundation classes), Atari was programmed using GEM, and BeOS used the Be API. Applications developed for Apple products have used various toolkits, but since macOS X it's called Cocoa (with desktops using AppKit and mobile devices using UIKit). Android devices are programmed with their own toolkit and other mobile platforms have explored other options.

The Unix and Linux operating systems have a more complicated story. Although the Motif toolkit was one of the first, the fact that its design offers multiple choices has meant there is no one true *look* or library. In the 1980s, before Motif was created, there was the OpenLook project that aimed to provide a standard interface look and feel for Unix systems. Although there were many different designs and toolkits to choose from, the main contributors to Unix decided that unification would help it compete with Windows and other desktop platforms. And so, in 1993, they chose Motif for future development.

One of the common features of desktop environment design is that it is frequently updated, as you could see in the screenshots of Microsoft Windows earlier in this chapter. Whether due to changes in fashion or advances in usability, these changes are expected and the Motif system did not adapt to them, and so new projects were created as alternatives. Later in the 90s, the GTK+ and Qt projects were started and provided a more modern, polished-looking user interface. Also, the Java platform launched with **AWT** (**Abstract Widget Toolkit**) in 1995, all of which were not platform-specific, opening a new world of cross-platform GUI libraries.

Cross-platform toolkits

The toolkits mentioned in the earlier subsection were all developed for a specific platform. They evolved along with the operating system design and are often developed using the manufacturer's preferred programming language. These challenges make it difficult (if not impossible) to create a single app that will work on all platforms. For that reason, the move to create a cross-platform toolkit requires taking a different approach and so developers started to design a library that could be written independently of platform specifics in a language that could be compiled for any of the operating systems it supports.

When GTK+ and Qt were created in the mid-1990s, they chose C and C++ (an object-based language derived from C), respectively. Both languages had wide adoption across most operating systems and were in use with some other toolkits already, keeping the barrier of learning low. The Java approach, however, was broader—to create a whole new language that would work across all of these platforms and deliver a graphical toolkit built on top.

Operating system and computer manufacturers have market power to sway technologies, and as new languages became available, they are able to force developers in the same new directions (such as Apple moving to Swift, Microsoft to C#, and Google moving their apps to Dart). However, the large open source communities built around cross-platform technologies are generally loyal to the language it is built with and so don't typically embrace such large changes. Therefore, these projects can be left behind in some senses and can encourage developers to look in new directions, such as web technologies.

Hybrid apps

As discussed earlier in the chapter, the WWW offered an attractive platform for delivering applications to users on multiple operating systems as well as providing a way to build apps once and run them on any computer. A web browser offers a highly customizable canvas so that, using **Cascading Style Sheets (CSS)**, any **HyperText Markup Language (HTML)**-based application can be made to look like any design. This benefit brought a lot of popularity to web application development and even influenced some native toolkits to expand their theme capabilities to mimic this.

Websites, as described earlier in this chapter, were designed for information transfer—initially for read-only, and later for sending and editing data too. To get from this to a fully interactive application requires additional programming capabilities, and for this **JavaScript** is used. In the years since its creation, JavaScript has grown in popularity and complexity—there are now multiple package managers to handle the thousands of available packages that can be used in any JavaScript-powered app. Of these libraries, many are graphical toolkits for handling application interaction and layout, much like a traditional GUI toolkit. Of these, the most popular are currently React, Vue.js, and JQuery, though that list changes regularly.

When developing such JS based web-apps, the focus is on the user interface entirely (the *frontend*), whereas a full application may contain business logic and algorithms that may be part of the server infrastructure (the *backend*) of a web application. Historically, these separate parts of a complex application are created using different programming languages. This makes sense as there are different requirements for each area of a large infrastructure. However, for smaller applications, or to reduce technical complexity, it can be useful to use the same language for all parts—and so Node.js was created to support the JavaScript-based backend elements of an application as well.

Despite the benefits of distributing over the web, many companies still want to provide a traditional application that is downloaded and installed (the reasons are explored in the next section). To balance the speed of development and other benefits of web applications with the standard application package that developers are familiar with, a new breed of applications was created, nicknamed **hybrid apps**. These new apps are loaded in a standard container that loads the custom web application in a regular window like any other application on the system. Electron, Ionic, and React Native are all projects working in this space, offering a web-based app framework with varying levels of access to the system they are running on.

Alongside the evolution of graphical toolkits for cross-platform development, we cannot ignore the ubiquity of the web browser. Although it has the benefit of being present on most modern computers, it may not be the right solution for building your product – let's take a look at how these approaches differ.

Comparing native graphical apps to web UIs

Despite the benefits that web-based applications can deliver, every technology choice means making a trade-off in some area, so let's look at a few of the common issues that might influence your decision of whether to build a native app or web-based hybrid.

Development speed versus delivery

One of the main reasons to pick a web technology to build your application is the speed of development. The very nature of developing in this way means that you can live-preview your work in a web browser. The availability of browser-based editors also means that a design team can tweak the user interface without much code experience. Large portions of your web app could also be used in your hybrid application (or the other way round) to provide a high level of reuse and minimal additional work to support desktop and mobile delivery.

The trade-off regarding speed is found at runtime—as the applications require a web view to run the code, there is an impact on how fast the application can be loaded and executed. Each time a hybrid app is loaded, it creates a small version of a web browser inside the window, loading the code like a web page and starting the execution of the bundled JavaScript. To most users, this may not be so slow as to be frustrating, but when compared to natively-compiled applications, there can be a noticeable difference. Depending on the framework chosen, it is also common for this model to require a lot of memory—indeed, Electron has a reputation for requiring a lot of RAM, with the simplest application using nearly 70 MB just to show 'Hello World'.

The actual speed of execution may also be noticeably slower for applications built on web technologies. Due to the layers of abstraction, a web-based app will typically take more time and CPU cycles to perform the same operations than a compiled native application (though technologies such as **WebGL** and **WASM** (short for **Web Assembly**) are attempting to improve this). Therefore, if your application is likely to be CPU-intensive, or have lots of animated graphics, you may wish to benchmark different approaches to determine which platforms are capable of meeting your requirements for app responsiveness.

Another consideration may be automatic updates—would you like your apps to always be running the latest version? Some web-based toolkits offer the functionality to download application updates and dynamically load the new version without the user having to worry. This can be a large benefit, but could also be frustrating if your customers expect the software to work exactly the same every day until they opt to update it. Some people are also concerned about how applications of this nature may appear to *phone home*—that is, reporting back to a central server about how the apps are being used, and where, as part of the update process.

Visual style

Another main decision point for choosing web technology-based app development may be the power of the presentation layer (CSS). It is possible, using a combination of image assets and style-sheet code, to create almost any visual style desired. For developers (or indeed, the designers on their team) that desire a completely bespoke look to their application, this could be a great fit. It is worth considering how your users will use the application, however, and whether a completely custom look will inhibit the usability in any way.

This benefit of complete customization can become challenging if the application is intended to match the user interface style of the current system. As the rendering is infinitely flexible, developers can of course add tweaks to a style that make it look subtly (or substantially) different when running on certain systems. This type of adjustment can end up taking a surprising amount of additional effort—as each platform can have a different style over time. A GUI that attempts to match the system style but does not quite manage this is far more off-putting than the one that is clearly following its own style guide.

Therefore, it is probably best to avoid hybrid apps if you desire to blend in with the other apps on a system. Web technologies do offer a fast-to-develop, adaptable platform for cross-platform applications, but there are constraints to this approach that should be considered as well.

Technical constraints

Applications built with web technologies, even those built to look like system apps with hybrid frameworks, run in a sandbox. This means that they are limited in certain ways regarding access to devices and system features. The JavaScript APIs that grant access to underlying functionalities are constantly expanding to work around these constraints, but if your application would benefit from non-standard peripherals or integration into specific operating system features, then a web UI may not be the right choice for you.

Access to communications ports, peripheral devices not included in typical web APIs, and process management are some low-level elements of an application that would not be supported by default. Additionally, interacting with a desktop environment's system tray, search features, and some advanced file management may be difficult to access from within the JavaScript code. To attempt to bridge this gap, some hybrid toolkits allow native code to be written and loaded as libraries to access this functionality. Such extensions, however, would need to be written in the platform's own language (usually C or C++) and then compiled for each supported platform. Not only does this increase the complexity of application distribution; it can also detract from the single-codebase app design technique that using web tools offers.

In comparison to this, other approaches to cross-platform development typically provide an abstraction over all supported operating systems so that an app can be built just once, but where support is missing, they provide a way to get direct access to the underlying features. This is often in the form of a language bridge or a way to load system libraries from the higher-level language. This can involve needing to build with different programming languages, as discussed earlier, but with cross-platform technologies outside of the web sandbox, it does not normally increase the distribution complexity as much. Additionally, it is far rarer to find devices not supported by a native cross-platform toolkit in comparison to one that is based in an embedded web browser.

If some of the constraints of web technology cross-platform development mentioned in this section might impact your app, or if you would prefer not to be coding with HTML and JavaScript, then native toolkits are probably the right approach, which we will look at next.

Options for cross-platform native toolkits

As indicated earlier in this chapter, the idea of a cross-platform toolkit is not new—in fact, they date back to the mid-1990s, barely 10 years into the history of GUI development. It is important to understand that even among native cross-platform toolkits there are distinct approaches with different benefits and drawbacks.

Visual style

Cross-platform toolkits can be divided into two distinct visual approaches—the desire to match the look and feel of the system at runtime versus the delivery of a distinct look that will be consistent across all the environments. The Qt and GTK+ toolkits began with their own visual style that added the ability to be controlled by visual themes. Over time, they developed operating system-specific themes that allowed them to match the design of other applications on a system. In contrast, the Java AWT library was created as a code-level abstraction, meaning that programs would use operating system widgets to render despite the application being written for no specific platform. Interestingly, in 1998, Sun (the creators of Java) introduced the Swing toolkit, which offered a whole new look and feel that would be consistent across all platforms. This replacement user interface library slowly gained popularity as AWT was phased out. In an interesting twist, Sun introduced operating system lookalike themes that a developer could choose to enable in their app (unlike in GTK+ and Qt, this was not a user preference).

Compiled versus interpreted

The other factor that commonly defines a toolkit is the choice of programming language—these can be split between those that are compiled into an application binary versus those distributed as source code that is interpreted at runtime. This distinction commonly correlates with statically typed languages compared to dynamically typed languages, which impacts programming styles and the design of associated APIs. In a compiled language that uses static typing, all variables are defined and checked at compile time. Their types (that is, the sort of content they contain or refer to) are set in the definition and never change. This approach typically catches programming errors early and can lead to robust code, but can be criticized as leading to slower application development. A compiled application will usually work directly on the computer it is built for, meaning that it does not require the installation of supporting technologies alongside the application. To achieve this, however, the apps will typically need to be compiled for each supported platform—again leading to longer development times required, this time for distribution tasks.

In comparison, an interpreted language is typically held up as supporting faster development and quicker delivery. By allowing variables to be used for different types of content, the source code of an application can be shorter with fewer complications during the programming phase. Conversely, the reduced rigor means that it is usually required to have a more solid test infrastructure (through unit tests or automated user interface testing) that can ensure the correctness of the software. Distributing apps in this way will require a runtime environment to be installed so that the code can be executed (like the embedded web browser we saw when discussing hybrid apps). For some operating systems, there may be interpreters installed already, and for others, they may have to be installed by the user. Interestingly, there are some compilers available for interpreted languages that allow them to be distributed like any other binary application, though with the preceding trade-off that applications must be built for each target platform individually.

Interpreted options

Due to the popularity of building GUIs, each of the main interpreted languages has its preferred toolkit. The Java runtime includes its own graphics routines, which means it can include a bespoke user interface, namely Swing. More recently, the JavaFX library has been built on top of the same graphical code. As mentioned earlier, Java also includes the AWT library, which delegates to system components.

Other popular programming language runtimes do not ship with the same graphical components and so they typically rely on an underlying library. TkInter is the standard GUI library for Python, based on the Tk library, while the Ruby language does not recommend a standard library. Language bindings, which allow programmers to create applications using an interpreted language while using an underlying existing widget toolkit, such as Tk, GTK+, or Qt, are very popular. Using this method, there are too many options available to list, but they usually have the same drawback of not being designed for the language and so can be less intuitive to program than options built specifically for the language being used.

Compiled options (C-based)

As shown earlier in this chapter, GUI development has a long history and the most popular toolkits were initially designed many years ago. GTK+ and Qt are both exceptionally popular, but they were designed for the C and C++ languages, respectively. While this does not stop them from being effective choices, it can make them seem a little outdated to the modern programmer. Possibly due in part to this reason, these GUI frameworks have seen language bindings made available for almost every programming language that exists. However, this does mean that you need to know a little about how the underlying system works to be most effective in development. For example, memory management in a C library is a complex and manual task, and not one that most developers want to worry about. Additionally, a C++-based library will have a different threading model beneath the surface that may not interact well with the higher-level language methodologies.

An additional consideration for toolkits that were designed decades ago is that they may not be best suited for the modern landscape of graphical computing devices. Screens, whether on a desktop, laptop, or mobile device, now come in a wide variety of sizes and pixel densities. This can bring challenges if code was built assuming that a pixel is a certain size, as used to be true. 96 **dots per inch** (**DPI**) was a common assumption, meaning that something 96 pixels high would measure about an inch when displayed. However, based on current devices, it could be anywhere from an inch down to 3/16ths of an inch (only 20% of the intended size), so it is important to understand how this will impact your application design. Older toolkits are typically based on pixel measurements, where the display is simplified to be a certain number of pixels (1, 2, 3, or 4) to represent each source pixel's size (which would be 1, 4, 9, or 16 pixels to display a square). Using a toolkit that scales in this manner can result in pixelated output if the application does not adapt carefully to the device.

Compiled options (other languages)

In part due to the legacy of the older C/C++-based GUI toolkits, but also in part because new programming languages bring benefits over the old ones, most of the newer compiled languages also have a toolkit built alongside. These more modern toolkits typically handle today's variety of devices much more effectively due to their more recent design. The challenges with pixel- or bitmap-based design is being overcome by moving to a vector image-based design. Vector images can be displayed at a higher quality output on screens of any pixel density. They do this by defining image elements such as lines, rectangles, and curves independently of output pixels; they are only drawn after the number of pixels available has been determined, resulting in higher quality on most output devices.

As we saw earlier, an operating system manufacturer often dictates the programming language used, and can push for new versions or programming languages when required. We can see this with Apple's recent move to Swift for the latest version of their UIKit and AppKit frameworks (this language is intended primarily for Apple devices but the trend is still noteworthy). Microsoft's development platform is currently geared toward C#, having previously used C, C++, and Visual Basic. Google is putting its energy into the Dart programming language and the Flutter toolkit built on top.

Other languages, such as Go, do not have an official GUI toolkit or widget library. In these situations, we can see various projects emerge with different ways of addressing the gap. In Go, the most active projects are andlabs UI, Fyne, and Gio. The andlabs project aims to use the current system style—in fact, it wraps the code to display them with a simple Go API—much like Java AWT discussed earlier. Gio is an immediate-mode GUI toolkit that aims to provide as much control as possible to the application developer, requiring the app to manage the rendering and event processing. The Fyne project aims for an API that is simple to learn and extend, without needing to worry about the rendering process—this is commonly known as *retained mode* as the widget state is managed by the library.

Summary

In this chapter, we explored the history of graphical applications and the toolkits that power them. We saw that over the last half-century, many things have changed, and yet a lot of aspects have stayed the same. Through an illustration of trends in graphical design and technical capability, it became clear that these technologies, while adapting and improving in the eyes of the end user, can be slow to embrace the speed of improvement that developers expect. We saw that there are many different approaches to supporting the creation of GUIs that work across multiple platforms, but also that they can have drawbacks as well.

In the next chapter, we will learn more about the Fyne toolkit's vision and design, and why the project team believes this provides the easiest way to build robust and performant graphical apps for any platform.

2

The Future According to Fyne

The Fyne toolkit design is based on the premise that the best way to resolve many of the challenges raised in *Chapter 1, A Brief History of GUI Toolkits and Cross-Platform Development*, is to take a fresh approach to GUI toolkit design. It aims to combine the benefits of a modern programming language, the Material Design look, and a simple API.

In this chapter, we explore the background and ambition of the Fyne project, including the following:

- The vision for Fyne and its team
- How does a modern programming language enable a fresh approach?
- How does it address the complexities of cross-platform, native app development?

Technical requirements

In this chapter, we will be using examples of Go code, so you will need to install the Go compiler – refer to the instructions at `https://golang.org/doc/install`. We will also explore bridging to C APIs, so you will need to install a C compiler as well. The installation of C varies from system to system. You can find details in *Appendix A, Developer Tool Installation*.

The full source code for this chapter can be found at `https://github.com/PacktPublishing/Building-Cross-Platform-GUI-Applications-with-Fyne/tree/master/Chapter02`.

Starting with a clean slate

Through the history of GUI development, we see that the majority of the most popular toolkits are based on C or C++ language code. These projects have a substantial history, large communities, and innumerable hours of development to make them what they are today. Despite being the yardstick against which all other toolkits are measured, they have drawbacks, mostly due to the legacy of the old design decisions they are built on. In this section, we reflect on why starting from scratch creates a better experience for building cross-platform apps.

Designing for modern devices

The types of devices we use today are both vastly different and much more varied compared with the 1980s and 1990s, when the most common toolkits were being designed and built. Today, a graphical application could be running on a desktop computer, laptop or netbook, a mobile device or tablet, smart phone or watch form factor, or even a set-top box or a smart TV. These device categories all have different user interface paradigms—APIs designed for a desktop computer using a mouse and keyboard don't always adapt well to touchscreen-based input, multiple touch hand gestures, or a remote control as the primary input device. An input API designed for modern devices will be able to provide a suitable abstraction over the low-level details to focus on the user's intent. By using this approach, applications can better adapt to the variety of devices in consumers' hands.

Along with the variety of device sizes come a wide spectrum of screen sizes and types. Mobile phones now have larger screens than ever, but they are still relatively small. However, the number of pixels they contain is huge, creating a smooth look when held close to our faces. TVs, on the other hand, are very large, but the resolution is low due to the distance they are usually observed at. Even desktop screens have changed—the average size has increased and the screens have become wider, but the biggest change is that pixel density has gone up astronomically, with high-end displays containing ten times as many pixels as those of cheap devices. Handling this variety of output devices requires resolution-independent rendering, which is a big shift from the origins of graphical toolkits, where pixels could be assumed to be a certain size. To work around these issues, some toolkits simply use a multiplier for each output pixel. This approach manages to avoid an app becoming too difficult to read by using more pixels to display the same original resolution—but will require application updates to avoid badly pixelated rendering.

With the ever-increasing complexity of smart phones, consumers now expect that applications will adapt to their location, preferences, and behaviors. Much of this is possible due to the inclusion of many sensors, each of which have platform-specific APIs. An easy-to-use abstraction over the platform specifics is part of a good modern toolkit and significantly reduces the time spent by developers in preparing their software for the different platforms.

Parallelism and web services

Computing technology has come a long way in 50 years, not just in machine rooms and server racks, but on our desks and in our pockets. It is often said that we now have more power in our pockets than we had when man first landed on the moon. In fact, it's more like 10,000 times the amount of power if you factor in a smart phone's capabilities. To be able to accommodate this much power requires modern programming techniques—it's not as simple as running the same code much faster. One major factor is that a modern operating system and any application on it will be able to calculate many things at the same time – but to do so effectively requires code that understands **parallel processing.** To make this adjustment is straightforward; code must be split into components that are independent enough to be able to run simultaneously. Introducing this ability can mean that memory is no longer controlled by a single part of code and so unexpected outcomes can occur (often called **race conditions**). Working around these complications with older tools can be very complicated and error prone.

Almost all GUI toolkits to date (including the latest releases by Apple and Microsoft) require that changes to the user interface or output graphics are handled by the *main thread*, a single portion of the capability of any application. This constraint requires careful coordination from an application developer and potentially limits the computing power available for the graphical components of an app.

One of the most common uses for background threads in an application will be communicating with remote resources. Web services or network resources are not accessible as fast as data on the local computer or bundled with an application, and so an app must manage user input and graphical updates while handling these slower requests. The core of most graphical toolkits and APIs is focused solely on the widgets— the presentation of the interface to the user. The design of many toolkits pre-dates cloud services and web-based APIs as we know them today. Powerful web services and standardized protocols for communication vastly improve the speed of development for web-based applications. Conversely, they can make it harder for native graphical applications on the desktop, where support is lacking from the core language or standard libraries. Inclusion of these features in modern programming languages is changing this and a modern graphical toolkit should provide similar benefits to developers.

Building for any device

While considering the hardware characteristics and connected nature of devices, we need to consider how software is deployed to them. Each operating system expects applications to be packaged and installed in a different format and often have different ways of discovering and downloading software as well. Although the C and C++ languages do work across most platforms, it can be very complicated to compile for a computer that is different to the one that you are working on. This restriction has not been a problem for desktop and laptop apps, as companies simply buy the different types of computers to run the compilation and packaging. It is, however, more complex for mobile and embedded devices, where they cannot compile their own apps from the source code.

Programming languages designed more recently include the ability to build for different systems from a single development computer. Although this is something that interpreted languages had done forever, it is now possible for your compiled applications (faster with no need for pre-installed runtime environments) to be **cross-compiled** for any device. In addition to this capability, the GUI toolkit must support preparation of different formats of application bundles that can package up the binary assets, resources, and other metadata for distribution. It is possible to perform these tasks manually, but a well-thought-out developer experience should ensure that this is practically automatic.

Additionally, many distribution platforms such as app stores and software marketplaces require a certification process that ensures authenticity of the apps it will make available. The cryptographic steps required to set this up are often lengthy and can be a hurdle to new app developers looking to make their work available. Rethinking a toolkit and the utilities it provides would allow this problem to be addressed alongside the compilation and packaging challenges.

Best practices have moved on

A final factor that can be observed when evaluating a toolkit for developing your app might be how up to date its best practices are. It has long been thought that automated testing of graphical applications is next to impossible, an opinion that is in no small way due to the fact that **Test-Driven Development** (**TDD**) and **Continuous Integration** (**CI**) were not common practices when the legacy toolkits, or the programming languages they use, were designed. A developer looking to learn, or one that is in a team of professional software engineers, would probably expect such practices to be supported or even encouraged in modern tools.

The C language (and many derivatives thereof) have often been criticized for the weak handling of string types. In fact, it is this very deficiency that has led to many highly visible software vulnerabilities and public data breaches. Although not all older programming languages suffer from this issue, they almost all have the restriction of only supporting simple strings using the *Latin alphabet* (although some have add-on libraries that attempt to work around this). With such a restriction, it is difficult to write an application that will easily adapt to the variety of commonplace languages that software should support in this international world. The **Unicode** standard is the universal approach to handling internationalized text, but this *multi-byte* (using more than a single byte to represent a letter or symbol) format can cause problems when introduced to software that was not designed to understand it. Users and developers alike now expect these complex encodings to be supported, and so the drawbacks of older toolkits continue to grow.

As you can see, there are a number of challenges in developing, or working with, a graphical toolkit that could be overcome if we start from scratch. And so, the Fyne toolkit decided to do just that, but to do so required a choice as to which programming language to use. As we know, Go was picked for Fyne—in the next section, we look at why it was seen to be a good fit to overcome the challenges faced.

How Go is a great fit for this challenge

In the previous section, we saw that there are many reasons why graphical toolkits are rooted in dated foundations and that even the languages they are built with could be limiting their chances of adapting. A number of manufacturers are recognizing this problem and reaching to new languages to find solutions or even avoid the challenges of the past completely. Apple is moving all development to the Swift language, although Apple-supported software is designed to run on their devices only. Other companies, such as Facebook, are finding ways to adapt more modern web-based tools to create native apps for phones and desktop.

Neither the approach of a platform-specific technology nor languages derived from the interpreted internet technologies are going to be able to truly create a delightful development experience. We are looking for a development platform that results in performant and robust, cross-platform applications—the panacea of modern application development. I, the author of this book and indeed the Fyne project, believe that Go may be the language to underpin such a revolution in building cross-platform graphical applications.

To quote the Go frequently asked questions on this topic, refer to `https://golang.org/doc/faq`.

> *Go addressed these issues by attempting to combine the ease of*
> *programming of an interpreted, dynamically typed language with the*
> *efficiency and safety of a statically typed, compiled language. It also aimed*
> *to be modern, with support for networked and multicore computing.*

In this section, we will look at the various reasons why the Go programming language is well placed to support a new era of GUI programming.

Simple cross-platform code

Go is a language that (like C, C++, Swift, and many others) compiles to a native binary on every platform it supports. This is important for graphical applications as it's the best way to create the most responsive and smoothest user interfaces on mainstream computer hardware. What stands out about Go compared to other languages that are popular with GUI developers is that it manages to support a long list of operating systems while compiling, without any alterations or special adaptation, to native code on every platform. This means that a Go-based project can be built on any computer for any other computer, using the standard tools, with no need for complex build configurations or extra developer packages to be installed. At the time of writing, the platforms that Go runs on includes Windows, macOS, Linux, Solaris, and other popular Unix-based operating systems (which is essentially all desktop personal computers) along with iOS, Android, and other Linux-based mobile devices (and even tiny embedded computers via TinyGo).

Go is a *typed* language, which means that every variable, constant, function parameter, and return type must have a single, defined type—leading to robust code by default. Unlike some older typed languages, Go is often able to infer a type, which helps avoid the duplication of information in the source code (in fact, one of Go's design principles is to avoid duplication). These features help to create a language that's fast and fun to develop with, while creating software that is as solid as the languages traditionally used for native graphical apps.

In addition to being easy to learn and simple to read, the Go language comes with well communicated standards for code style, documentation, and testing. Standardization such as this makes it easy for developers to understand different projects and reduces the time required to integrate libraries and learn new APIs. As well as documenting these standards, the Go development tools include utilities that can check whether your code meets these guidelines. In many cases, they can even update your source code files to comply automatically. Quite naturally, the development environments that support Go also encourage following the guidelines, making it even easier to reduce the learning curve for anyone joining your project.

As well as standard formats for source code and documentation for all APIs, the Go language and community support and encourage unit testing in apps and libraries. The compiler has built-in test functionality normally associated with *dynamic* languages that need this sort of check to ensure correctness. The inclusion of effective testing alongside an already robust language provides validation of code behavior and makes it easier for code to be changed by individuals other than those that created it. In fact, in a collection of popular Go libraries, one of the requirements for being listed is that your code meets an 80% unit test code coverage metric (see `https://github.com/avelino/awesome-go` and their contribution guidelines).

Standard library

The standard library of a programming language is the set of APIs and features that are provided by the language runtime. C, for example, has a very small standard library—as a low-level language designed for many different types of devices, the number of features that it can support for every operating system is limited. Java, on the other hand, historically known for being heavy on memory and start up time, provides a massive standard library—including the Swing GUI described in *Chapter 1, A Brief History of GUI Toolkits and Cross-Platform Development*. This is a trade-off that all languages need to make when deciding in favor of smaller memory or lots of built-in features.

Thankfully, the Go language makes a clever balance, allowing it to comprise a large library of APIs that fully support every one of its target operating systems. To do this, it makes use of **build tags** that allow inclusion of only the code needed for the current (or target) operating system. This is a huge advantage for developers who want to write an efficient application for multiple operating systems without maintaining slightly different versions for each platform, or suffering slow load times or large memory requirements.

The standard library included with Go includes powerful features across many areas, including cryptography, image manipulation, text handling (including Unicode), networking, concurrency, and web service integration (we will cover this in the sections to come). You can read the full documentation at `https://golang.org/pkg/#stdlib`.

Concurrency

As illustrated in the *Parallelism and web services* section earlier, a modern programming language needs to handle concurrency. Unfortunately, working with the APIs to manage multithreading can add complexity and make code harder to read. The designers of Go decided that concurrency should be incorporated from the beginning, making it easy to manage many threads of execution while still avoiding the difficulty of shared memory management. GUI toolkits built on languages without this built-in awareness of concurrency have propagated the idea that graphical routines must happen on a particular thread. By starting from scratch with a better suited language, we can avoid such constraints.

Go does not expose traditional threads, but instead introduces the concept of **goroutines**—these are akin to lightweight threads, but that are enhanced to support several thousand at the same time. It is common for applications to communicate between background processes by sharing memory, but this introduces issues known as *race conditions*, requiring more code to manage access. To avoid this, Go provides **channels**—a mechanism for communicating between threads of execution without causing the same problems. With this model, the language manages the safe transfer of information from one goroutine to another, keeping multithread code neat and easy to understand.

Web services

As a modern programming language, Go comes with extensive support for HTTP clients, servers, and standard encoding handlers, including JSON and XML. For GUI developers coming from a background in C programming, this is a significant improvement—web services, and indeed remote resources, were not commonplace when the language or its toolkits were created.

Thanks to language-level support for text encoding used in web communications, it is possible to load data structures directly from an HTTP request. Such convenience may be standard for web-based languages such as JavaScript and PHP, but to function in a strictly typed language without using third-party code is a rare bonus.

Despite these features that make Go an excellent language for building complex applications that can be described at a high level and thoroughly tested, it is also possible to perform platform-specific requests to the operating system if required.

High level with system access

Even with a fully featured programming language, it can be occasionally necessary to access low-level components or platform-specific APIs. Whether it is to send a notification, read data from a custom device, or simply to call a function specific to the current operating system, it will occasionally be necessary to access functionality that is not included in a programming language or its standard library. To solve this challenge, Go provides three avenues to handle the platform-specific aspects of a program—syscall, CGo, and build tags.

Build tags

In Go, it is possible to include part of your code for specific operating systems at a per-file level based on some conditional parameters called **build tags**. This can be helpful to adjust behavior when your app will run on certain systems, but is more beneficial in controlling the inclusion of the platform-specific use of syscall or CGo code illustrated in the next sections.

Even if your app is not making specific use of operating system-specific calls, it can be helpful to make use of the conditional compilation. In simple cases, Go code can check what system an app is running on to execute slightly different code (by checking the value of os.GOOS, for example, a function could return different values). However, for more behavioral changes, it can be useful to put the platform-specific code into separate files that are named in a certain convention (for example, `*_windows.go` will be included when building for Microsoft Windows) or through the use of the special comment at the top of a file (such as `// +build linux,darwin` including the file when Linux or macOS is the target platform).

Syscall

One of the packages provided by Go's standard library is **syscall**. This package provides a way to access the low-level, operating system-specific functionality, for example, accessing network sockets or managing desktop windows. The functionality that is available will vary based on the underlying system (the target operating system), but can be very powerful when combined with build tags. This package is used by other built-in packages, such as os and net, to provide higher-level abstractions, so double-check that the functionality you desire is not provided by another package before using syscall.

A typical call into this package might be to request details of a Windows registry key, or to load a DLL (a system library on a Windows computer) to access functionality not provided by Go. On a Linux computer, you may use this functionality to read or write to specific memory areas such as connected devices, where permitted. A system call is a very complicated procedure and, where possible, it will normally be easier to call a C function instead – we see how to do that with CGo next.

CGo

If your application, or one of the libraries that you rely upon, was written with the C language (or one of its derivatives, such as C++ or Objective-C) for specific reasons and you're not able to move it to Go, then CGo is an invaluable feature. Using this functionality, it is possible to include C code directly, or call out to other C-based functionality. It is important to use this carefully—there are some performance impacts in making the jump from one language to another—but more significantly you need to remember about memory management and threading in a way that Go would normally manage for you.

The following source code shows C and Go code in the same file where we convert a Go string into a C string (or, more strictly, a [] byte) and pass it into a C function. As you can see, we also need to free the memory we passed into the C code once it is no longer required:

```go
package main

/*
#include <stdio.h>
#include <stdlib.h>
void print_hello(const char *name) {
    printf("Hello %s!\n", name);
}
*/
import "C"
import "unsafe"

func main() {
    cName := C.CString("World")
    C.print_hello(cName)
    C.free(unsafe.Pointer(cName))
}
```

The preceding code can be run just like any other Go program, as follows:

```
$ go run cgo.go
Hello World!
$
```

In the code snippet, you can see that the C code is included in a comment above the `import "C"` line. This code could be in a separate `.c` file or even within a library on the build computer (and the Go compiler will use `pkgconfig` to find the required headers). When used in combination with conditional building such as build tags in the previous code, you can see that it is possible to access platform-specific functionality or legacy code where required.

This section considered the Go language and how its design is well suited to building a modern GUI toolkit. Next, we will look at how Material Design provides a great aesthetic for Fyne applications.

Looking good with Material Design

A key part of any GUI toolkit that can impact the developer's selection and also the user appeal of the applications built is the overall design language. This choice of aesthetic can be seen in the colors, fonts, layouts, and even icon design. Some of these choices are obvious and others more subtle, but, when combined, result in a recognizable application look and feel.

New cross-platform toolkits commonly create their own design, such as Java's Swing or the GTK+ and Qt toolkits. These are often designed to look contemporary with the software of the time—you can recognize a 1990s desktop application design in those toolkits. In the current landscape, usability and design principles of mobile apps are being adapted and deployed to other areas, bringing a new age of software design to traditional applications. Because of its work in this space, the Material Design project makes a good match for application design aiming at universal cross-platform appeal.

The official website for Material Design (`https://material.io`) describes material as follows:

> *Material is an adaptable system of guidelines, components, and tools that support the best practices of user interface design. Backed by open source code, Material streamlines collaboration between designers and developers, and helps teams quickly build beautiful products.*

Google first published the Material Design guidelines in 2014 based on their earlier work on a new design language for their websites. It has since been applied to their different web properties and has also become the visual design for the Android operating system. This adaptation meant that the component designs have identified areas that work across all platforms and also where slight adaptation may be required for desktop computers, reducing the work required by Fyne or other cross-platform toolkits that use its design language.

The layout and functionality of the standard components varies by platform, and implementation though clear recommendations is documented at the main website, `https://material.io`. What can be more universally recognized is the color palette and iconography used in material-based apps.

Color palette

Material Design colors are oriented around a standard color palette. A primary color is used to raise the importance of certain elements, such as default buttons or focused input. A complementary secondary color is (optionally) reserved for accenting important items, such as floating buttons or selected text. The Material Design project provides a baseline color theme that can be used with any app. Developers may also choose their own primary and secondary colors to match their brand identity or preferred aesthetic. Here is the baseline color palette:

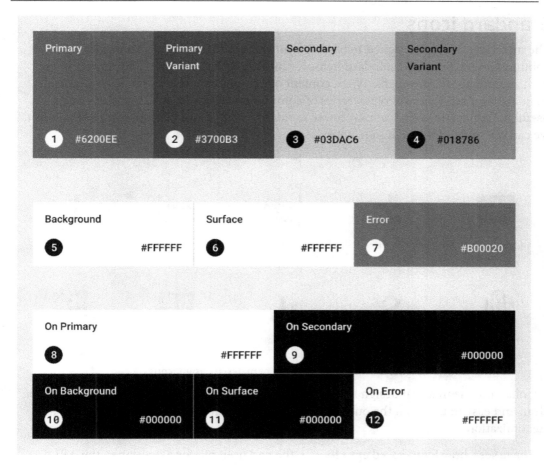

Figure 2.1 – The material design "baseline" color palette

The striking colors of a material theme help an application to have a clean design while conveying meaning and importance. Similarly, the material design offers a set of clean, crisp icons that should be used where possible.

Standard icons

The material icons are designed to be *delightful and beautifully crafted* and cover the common interactions and standard items for most modern software. They cover topics such as standard hardware, file types, content operations, and user actions. As well as the standard icon set, there are repositories of community submitted add-ons that can be useful for applications that employ fewer standard operations. In the following screenshot, we can see some material design icons:

Figure 2.2 – A small selection of material design icons

By following the material design specification, any apps built using the Fyne toolkit will be clean and easy to use from the outset, while supporting an element of brand identity and customization.

We have seen how Fyne-based apps look to the end user, but for a developer, the API design is just as important. Let's look at how the team aim to keep that just as clean and well crafted.

Designing APIs for simplicity and maintainability

A delightful user experience is an important ambition for any application toolkit, but Fyne aims to make the experience of development a pleasure as well. To do this, the API must be planned carefully to be simple and easy to learn, but also extensible to support more complex applications. The project's modular approach aims to support this while also being testable every step of the way.

Semantic API

An **API** (or **Application Programming Interface**) is typically defined as a set of functions and procedures that controls access to features and data. At a high level, however, the Fyne toolkit aims to deliver a *semantic API*, a set of functions that defines intent rather than features or functionality. By taking this approach, the toolkit is able to separate meaning from presentation.

For example, we can consider a simple button—when there are many on a screen, you may wish for one to stand out against the other as more important. In an API that is focused on presentation or styles, you might set the button color; with the Flutter GUI toolkit, this would appear as follows:

```
FlatButton(color:Colors.cyan, child: Text("Tap Me"))
```

In comparison, an API that takes a semantic approach would allow developers to indicate the expected difference through a button type or intent field, as the following Fyne snippet does:

```
widget.Button{Text: "Tap Me", Importance: widget.
    HighImportance}
```

Following this approach allows for a consistent API that describes the expected outcomes instead of features that could hint at those outcomes. It also allows the current theme to ensure a consistent visual style and avoids developer-defined code creating unreadable or unattractive graphical choices.

Modular

When building a robust toolkit designed to work seamlessly across many different operating systems and computers, it is important to take a modular approach. Doing so ensures that no element of the library will be able to make false assumptions about other areas of the code. To accidentally expose all the inner details of a key area, such as a graphics driver, might restrict a widget to function correctly on a single operating system or in a particular graphics mode. This technique is known in software engineering as **Separation of Concerns**.

In the Go language, modules are known as **packages** and they are structured hierarchically under the project root. To allow different parts of the system to communicate, a project typically defines a collection of **interface** types that describe the functionality, and dependencies that some code may choose to conform to. By loading code that implements these interfaces, an application or library can compose separate elements of code to create the complete solution. Each area is only aware of the declared capability that each interface publicizes and can hide all of the inner details. This allows complex software to be built and tested in smaller pieces, which is far easier to test and even to debug if something goes wrong.

In Fyne, the use of packages can be seen in many areas, with the most notable being implementations of the `Driver` and `Widget` interface definitions. The use of drivers in the Fyne toolkit makes it possible for applications to run on many different types of computers without needing to know, or accidentally taking advantage of, specific details of a single device. When an application starts, the correct driver will be loaded to handle specific details of running on the current computer. As you will see in *Chapter 5, Widget Library and Themes*, the various widgets inside Fyne (and indeed custom ones that can be added by app developers) all implement the `Widget` interface. The behavior that all widgets must implement provides information to the driver and graphics code about how it should appear, meaning that the graphics code does not need to know any inner details of a widget to be able to draw it in an application window. This makes it possible for widget developers to avoid impacting graphics code, and indeed for other developers to add custom widgets outside the toolkit code.

One other benefit of a modular approach is that code can be executed without launching in a standard application or showing any windows. This may not be a common requirement for user-facing code, but it is very important to support the efficient testing of applications.

Testable

Automated testing of graphical user interface code has long been considered one of the hardest steps of a full test suite, and this becomes even more difficult if you add smart phones or mobile devices to your supported platforms. Each operating system may require a different methodology and probably also specific code to be written and maintained in order to run the test suite. As mentioned in the previous section, the modular approach taken by the Fyne toolkit promises easier testing by not requiring applications to be displayed for the test scripts to execute.

Because of its modular design, the drawing components of a Fyne-based application are a minor detail. The main logic and behavior of widgets are defined completely separately to the graphical output. This approach allows all elements to be tested through automated interactions faster and more reliably than alternative tools that load applications and then tap buttons using test runners that control the mouse and keyboard (or touchscreen) hardware. The following code snippet shows how a Button and Entry widget could be tested in a simple unit test:

```
func TestButton_Tapped(t *testing.T) {
    tapped := false
    button := widget.NewButton("Hi", func() {
        tapped = true
    })

    test.Tap(button)
    if !tapped {
        t.Errorf("Button was not tapped")
    }
}

func TestEntry_Typed(t *testing.T) {
    entry := widget.NewEntry()
    test.Type(entry, "Hi")

    if entry.Text != "Hi" {
        t.Errorf("Text was not updated")
    }
}
```

As you can see, these two simple unit tests (using the standard Go test structure) are able to test that the Button and Entry widgets behave as expected when standard user interactions occur. The test helpers Tap and Type are provided to perform these actions, along with various other utilities in the test package. By building up a test suite in this manner, you can execute thousands of GUI tests per second, without ever having to load a window or connect to a specific device. In fact, this very functionality supports true TDD for graphical applications. This approach means that application code can be designed and understood before it is implemented, leading to more robust software and better decoupling of modules, thereby allowing more developers to work on a project concurrently.

The Fyne toolkit will ensure that all elements will be presented correctly on any device that your app will be distributed to. Its drivers and widgets all undergo the same testing rigor described earlier in this section. Sometimes, however, it is necessary to test the actual rendered output of a widget or screen. In this situation, the `test` package has more utilities that can aid your development. Although not visible on your screen, the Fyne test code will calculate how the output would look and can save this to an image through a `Capture()` function. The test helper `AssertImageMatches` is then able to compare this to a specific output either saved earlier or created by a designer:

```
func TestButton_Render(t *testing.T) {
    button := widget.NewButton("Hi", func() {})
    window := test.NewWindow(button)
    test.AssertImageMatches(t, "button.png", window.Canvas().
        Capture())
}
```

This code sample does include some details about `Window` and `Canvas` that we will cover in *Chapter 3*, *Windows, Canvas, and Drawing*, but you can see the overall simplicity. The code defines a widget (in this case a `Button` widget), and then adds it to a test window that is then captured and compared to a pre-existing image file. The test window is not shown on screen or even communicated to the operating system—it is loaded purely in memory to simulate the drawing process.

We have seen how good modularity and testing can result in more robust applications, but this has described the toolkit design—is it possible for developers to extend the functionality for their own purposes?

Extensible

The core widgets of the Fyne toolkit are designed to be robust, easy to use, and well tested, but a toolkit cannot include every possible type of widget. For this reason, a toolkit also needs to be extensible, supporting the inclusion of widgets that were not defined by the core project—either as add-on libraries or by allowing applications to add their own custom user interface elements.

The Fyne project allows widgets to be included in two different ways—developers can extend existing widgets (keeping the main rendering consistent, but adding new functionality) or by adding their own widgets. As described in the *Modular* section earlier, any code that implements the `Widget` interface will be interpreted as an interface component and can be used throughout any Fyne application. Later in this book, we will also see how existing widgets can be extended to add new functionality or tweak behavior to suit a particular application.

Due to the interface-based design of the modular code base, there are many other ways in which a Fyne application can be extended. By implementing the `Layout` interface, an app can define its component position and sizing, or, using the URI interface, it could connect to different types of data resources (for more information, see *Chapter 4, Layout and File Handling*).

As you can see, the design of a toolkit and its API is just as important as the functionality it contains. To complete this chapter, let's review the overall vision of the Fyne project.

A vision of the future

The Fyne project was created in response to growing criticism of the complexity in existing graphical toolkits and application APIs and their inability to adapt to modern devices and best practices. It was designed with the aim of being easy to use, and the Go language was chosen for its powerful simplicity.

The vision statement on the Fyne project wiki (`https://github.com/fyne-io/fyne/wiki/Vision`) states the following:

> *Fyne's APIs aim to be the best for developing beautiful, usable, and lightweight applications for desktop and beyond.*

With more device types and platform-specific toolkits than we have seen in recent times, it is more difficult, and more expensive than ever, to deliver a great native app experience across all platforms. The Fyne toolkit is positioned as a solution to these challenges, while bringing the design and user experience learning from modern mobile apps to all devices.

Beautiful apps

Fyne aims to support the building of graphical applications that look consistent across all platforms, rather than adopting the look and feel of the operating system. Its APIs ensure that all apps provide a polished user experience and render a beautiful application GUI. Using the Material Design guidelines, a Fyne-based app will look familiar to Android users and match the aesthetic of the flat user interface in Windows 10. For operating systems where the user interface is typically in a different style, users will still be delighted by the crisp visual design, bundled icons, and clean typography.

As many operating systems now offer a choice of light and dark modes, toolkits need to adapt appropriately to meet user expectations. All Fyne apps include a light and dark theme and, unless the developer overrides the setting, it will match the current system configuration. When end users change their system theme, any running Fyne apps will update to reflect configuration changes. This is shown in the following image, we see what the light and dark Fyne themes look like:

Figure 2.3 – Fyne light theme
and dark theme, respectively

Simple to learn

As well as maintaining a clean, simple design, the Fyne team wants to ensure that everyone can learn to build graphical applications. To do this, the barrier to entry needs to be low—through simple installation and setup with documentation and tutorials readily available to support even the most inexperienced developer.

Developers who have not used the Go language before can start with the online tour (`https://tour.golang.org/`), and from there move to the Fyne tour, which introduces developers to the concepts of GUI development and how to get started with the project (`https://tour.fyne.io/`). Lots more documentation is available at the main developer website, containing tips on getting started, code tutorials, and a full API reference (`https://developer.fyne.io/`).

Platform-agnostic

Cross-platform toolkits have various levels of complexity when building for different platforms. Some require different build processes and others will load different user interfaces, depending on the type of device. Fyne aims to be easier to use across these disparate targets and aims for **platform-agnostic**, meaning that the code doesn't need to know anything about the device it is running on.

> *For many start-ups, the term platform agnostic represents a kind of unattainable utopia in the world of mobile apps.*

This quote is from `http://alleywatch.com`.

The code of a Fyne-based app does not need to adapt to system specifics and can compile for any devices using the same set of tools. As we will see later in the book, it is sometimes necessary to adjust for a specific operating system or device. In these situations, there are Go and Fyne APIs available to help. For many applications, it will be possible to avoid any customization per-system and the only element of the process that varies will be the application distribution.

Distribution to all platforms

When your app is ready for release, it needs to be packaged and uploaded to a central place where users can find it. Unfortunately, every operating system uses different packaging formats, and every vendor has a different store or marketplace for apps. The `fyne` command-line tool that we will use many times throughout this book is able to create application bundles in all of the required formats. With the app bundled, it can be installed locally, shared with friends, or uploaded to a website for distribution.

Most systems are now moving to an app store or marketplace model where apps are available in a manufacturer provided location with screenshots, free advertising, and a managed installation. One of the challenges here is that every store is different and the process of certifying and uploading apps is different for each platform. The Fyne tool helps with the release process as well—streamlining as much as possible and ensuring a consistent developer experience for all stores.

Summary

In this chapter, we have seen how designing a new graphical toolkit could overcome many of the challenges still faced by existing approaches. We explored the background and vision for Fyne, how it aims to solve these difficulties, and how it supports creating beautiful and performant apps across all popular desktop and mobile devices. We introduced Material Design and saw how it brings modern usability principles and design learnings to desktop and beyond. By using the `fyne` build tools, we saw that an app can be built and distributed for any devices or app stores without any platform-specific code.

In the next chapter, we will explore the fundamentals of a Fyne application and see how its drawing capabilities allow us to build a simple game.

Further reading

To learn more about some of the topics introduced in this chapter, you can visit the following websites:

- Test-driven development: `https://www.agilealliance.org/glossary/tdd/`

- The Go programming language: `https://golang.org/`

- C bindings from Go: `https://golang.org/cmd/cgo/`

- The Material Design system: `https://material.io`

- Fyne toolkit developer documentation: `https://developer.fyne.io/`

Section 2: Components of a Fyne App

The Fyne toolkit has many packages of components that build on each other. Providing a full cross-platform widget toolkit requires many enabling areas, including graphics **application programming interfaces (APIs)**, file handling, and layout management. In addition to this, Fyne provides advanced functionality such as data binding and storage abstractions that enable complex applications and workflows to be built with ease.

To understand the full capabilities of Fyne we will explore all the areas of the toolkit, its packages, and its functionality. Each of the chapters in this section includes an example application that explores the concepts, with detailed steps and full source code.

This section will cover the following topics:

- *Chapter 3, Canvas, Drawing, and Animation*
- *Chapter 4, Layout and File Handling*
- *Chapter 5, Widget Library and Themes*
- *Chapter 6, Data Binding and Storage*
- *Chapter 7, Building Custom Widgets and Themes*

We start this section by exploring the canvas and graphics APIs that power the whole toolkit.

3
Window, Canvas, and Drawing

We have explored the basics of graphical application development and seen how starting with a new design in a modern language can lead to easier development. From this point on, we will be looking in more detail at how the Fyne toolkit aims to provide an easy-to-use API for building cross-platform applications for all developers.

In this chapter, we will investigate the structure of a Fyne application, how it draws objects, and how they can be scaled and manipulated—as well as animated—in a container.

In this chapter, we will cover the following topics:

- How Fyne applications are structured and how to start making your first app
- Exploring the canvas package and the types of objects that can be drawn
- How scalable elements create a clean user interface
- Working with bitmaps and pixel rendering
- Animation of elements and properties

By the end of this chapter, you will know how these features combine to create a graphical application, which will be demonstrated with a simple game.

Technical requirements

In this chapter, we will be writing our first Fyne code, including building a complete application. To do this, you will need to have the **Go** compiler installed, as well as a code editor application. You can download Go from the home page at `https://golang.org/dl/`. The choice of code editor is normally a matter of user preference, but Microsoft's *Visual Studio Code* and JetBrain's *GoLand* applications are both highly recommended.

As Fyne uses some operating system APIs internally, you will also need to have a **C** compiler installed. Developers on Linux will probably already have one; macOS users can simply install *Xcode* from the Mac App Store. Windows-based developers will need to install a compiler, such as *MSYS2*, *TDM-GCC*, or *Cygwin*—more details can be found in *Appendix A – Developer Tool Installation*.

The full source code for this chapter can be found in the book's GitHub repository at `https://github.com/PacktPublishing/Building-Cross-Platform-GUI-Applications-with-Fyne/tree/master/Chapter03`.

Anatomy of a Fyne application

As we saw in *Chapter 2, The Future According to Fyne*, the toolkit took the opportunity to start from scratch, throwing away the old and sometimes confusing constraints of previous toolkits. As a result, the APIs need to define everything involved in building a graphical application. In this section, we will explore the main concepts in running a Fyne-based application and producing visible components on screen, starting with the application itself.

Application

The application, defined in the `fyne.App` interface, models the capabilities of a Fyne-based application. Each app using Fyne will typically create and run a single `fyne.App` instance from within their `main()` function. Because of the way that graphical applications work, they must be started from the main function and not through a goroutine or other background thread.

To create an app instance, we make use of the `app` package within Fyne, which can be imported using `fyne.io/fyne/v2/app`. This package contains all of the logic and driver setup code that allows an app to understand the platform it is running on and configure itself appropriately. The function that we call is named `New()` and it will return the app instance that we will use throughout our code. To run the application, we would then call `Run()` and the application will start.

With this knowledge, we could run our first Fyne app; however, it would be difficult to know if it was working without asking it to display something first! And so, we shall now learn how to display a window before running the first example.

Window

A **window** defines the area on the screen that your application controls. In a desktop environment, this will typically be shown inside a window border that matches the rest of the applications installed. You will normally be able to move and resize the window and close it when you are done.

On mobile and other devices, this concept can be a little less well defined. For example, on Android and iOS smart phones, the application window will take up the whole screen and will not show window borders. To switch applications, you would use a gesture defined by the operating system or press a standard button, and other applications will appear, allowing you to move around. In addition to this, tablet computers—iPadOS, Android, or Windows, for example—will allow applications to be displayed in a portion of the screen, probably separated by a divider that allows the user to change how much space is used.

In all of these different presentation modes the content that is displayed is still modeled by the `fyne.Window` interface.

To show content on the screen, we will need to create a new window and show it. Once a window is defined, we can run an application to see the results. Let's work through the code for our first application!

1. We open a new file, `main.go`, and define it to be in the package `main`:

```
package main
```

2. We then need to add any imports—in this case, we are just using the `app` package, so the following will be sufficient:

```
import "fyne.io/fyne/v2/app"
```

3. To define a runnable program, we create a `main()` method. Into this function, we will create a new application instance using the `New()` function we saw earlier:

```
func main() {
    a := app.New()

    ...
}
```

4. Also, in this method, we call `NewWindow(string)` (defined on `fyne.App`),
    which allows us to create the window to display. We pass it a single string parameter
    that sets a title (which is used if the operating system shows titles in the window
    border or app switcher, for example). Place the following code where . . . appeared
    in the previous snippet:

    ```
    w := a.NewWindow("Hello")
    ```

5. Once we have created a window, we can show it using the `Show()` function.
    After showing the window, we also need to call the `Run()` function on the
    application. As it is common to do both at the same time, there is a helper
    function `ShowAndRun()`, which we can use when showing the first window in an
    application:

    ```
    w.ShowAndRun()
    ```

6. With this code in place, we can save the file and run its contents like any other Go
    application:

    ```
    Chapter03/window$ go run main.go
    ```

7. You should see a window appear on screen. Depending on your operating system,
    this may be a very small window, as we have not added any content. The following
    screenshot was taken on a macOS computer after resizing the empty window:

Figure 3.1 – Our first window

As you can see from the window in *Figure 3.1*, there is no content to the window, as we did
not set any. The background of the window is not just black (or random colors from old
graphics memory)—how can this be? The reason is that the window contains a `Canvas`,
which is what manages the content we draw.

Canvas

The content of every `fyne.Window` is a `fyne.Canvas`. Although the way that the
canvas works internally is dependent upon the current system and some complex code
within Fyne's internal driver packages, it will look exactly the same to developers and our
application's end users. This platform-independent rendering canvas is responsible for
all of the draw operations that come together to create a graphical output and eventually
complete application interfaces.

Inside every canvas is at least one `fyne.CanvasObject`. These objects, as we will see in the *Understanding CanvasObject and the canvas package* section, define the types of operations that can be drawn to the canvas. To set the content of our window we can use `Window.SetContent(fyne.CanvasObject)`. This function passes the content down to the canvas, telling it to draw this object, and also resizes the window to be large enough to display it.

Of course, this only sets the content to be a single element; we will normally want to include many, and this is what the `Container` type provides.

Container

The `fyne.Container` extends the simple `fyne.CanvasObject` type to include managing multiple child objects. A container is responsible for controlling the size and position of each element it contains. *Figure 3.2* shows how the **Canvas** contains one container that positions three **CanvasObject** elements in a stack on the left and an additional **Container** to the right. This second container is responsible for three further elements, which it lays out in a row:

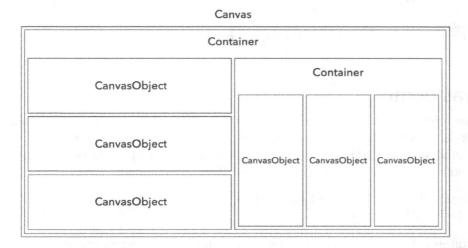

Figure 3.2 – A Canvas with various Container and CanvasObject elements

Containers typically delegate the work of positioning child objects to a `fyne.Layout`. These layout algorithms will be explored further in *Chapter 4, Layout and File Handling*. In the current chapter, we will use containers without layouts—these are called *manual layouts* and are invoked using `container.NewWithoutLayout(elements)`, where the `elements` parameter is a list of `fyne.CanvasObject` types that the container will present. We will explore manual layouts further in the *Combining elements* section.

Now that we have seen how the application is defined and how it handles the presentation of graphical elements, we should see what drawing capabilities Fyne supports, and how to use them.

Understanding CanvasObject and the canvas package

The `CanvasObject` definition is just a Go interface that describes an element that can be positioned, sized, and added to a Fyne canvas. The type does not contain any information about how to draw—this information is provided by *concrete types* within the `canvas` package. These types define well-understood graphical primitives, such as `Text` and `Line`.

Before learning how to use these elements, we shall see how they look in the Fyne demo app.

Canvas demo

Before we look at how to write code that will display shapes in our window, we should look at a demo of these features in action. Using the built-in Fyne demo application, we can see what the `canvas` package supports. If you have not already done so, you can install and run the demo application using the following commands:

```
$ go get fyne.io/fyne/v2/cmd/fyne_demo
$ ~/go/bin/fyne_demo
```

While running the demo, tap on the **Canvas** item on the left navigation panel. You should see the following screen:

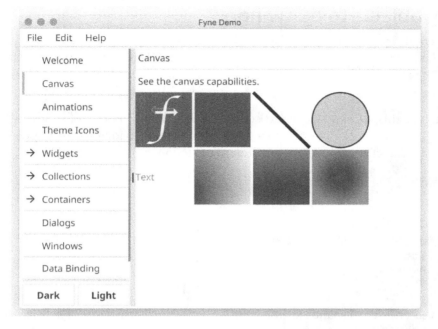

Figure 3.3 - The fyne_demo application showing various canvas primitives

The window shown in *Figure 3.3* demonstrates some of the canvas types known to Fyne. Those drawn here are named in the following list in order (left to right, from the top row and then the bottom):

- Image
- Rectangle
- Line
- Circle
- Text
- Raster
- Linear Gradient
- Radial Gradient

Each of these elements can be included in our application, as we will explore next.

Adding objects to our window

Each element in the previous demo figure, as well as any new items that are subsequently added to the `canvas` package, can be created directly using their `NewXxx()` constructor function. The object returned from this can then be passed directly to the window or a container of objects.

To demonstrate this, let's add some text content to the empty window. After adding `image/color` and `fyne.io/fyne/v2/canvas` to the previously used `import` statements, we can change the main function to the following:

```
func main() {
    a := app.New()

    w := a.NewWindow("Hello")
    w.SetContent(canvas.NewText("This works!",
        color.Black))
    w.ShowAndRun()
}
```

As you can see, this change adds just one line—the `canvas.NewText` that is passed to `w.SetContent`. The text constructor function takes two parameters, the text to display and the color to use. If you run this code, you will see that the window now contains the text **This works!**, and is sized just right for this to be displayed:

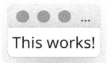

Figure 3.4 – Displaying text content

As you can see here, showing a canvas element is as simple as knowing which content you would like to display. Let's look at a slightly more complicated example using multiple canvas elements in a container.

Combining elements

To show how we can use a container to display multiple items and create a more appealing output, we shall replicate a road sign using a `canvas.Circle` and `canvas.Rectangle` element inside a `fyne.Container`. Let's see how to go about this:

1. Firstly, we will create a new function for this code, named `makeSign()`. It should return a `fyne.CanvasObject` (which all graphical elements implement). The rest of our code will go into this function:

```
func makeSign() fyne.CanvasObject {
    ...
}
```

2. Then we will create the background for a sign using `canvas.NewCircle()`, saving a reference to it so that we can use it later. The color that we pass is bright red—255 (the maximum) for red channel. Green and blue values are 0, and the alpha channel (how opaque the color appears) is also set at maximum, so it is fully visible:

```
bg := canvas.NewCircle(color.NRGBA{255, 0, 0, 255})
```

3. Then we add a white border to this circle. The `StrokeWidth` property controls how wide the border is (defaults to 0 or hidden) and we set `StrokeColor` to white for an outer circle of white:

```
bg.StrokeColor = color.White
bg.StrokeWidth = 5
```

4. Next, we will draw the bar across the center; this is simply a white rectangle:

```
bar := canvas.NewRectangle(color.White)
```

5. To combine these two elements, we define a container. It is important to pass the parameters with the circle (`bg`) first and rectangle (`bar`) second, as they will be drawn in this order:

```
c := container.NewWithoutLayout(bg, bar)
```

6. Next, we must position these elements. We will specify that the sign is 100 x 100. Because it has a border, we will inset it from the sides by 10 x 10 so that it is centered within a window sized 120 x 120:

```
bg.Resize(fyne.NewSize(100, 100))
bg.Move(fyne.NewPos(10, 10))
```

7. To position the bar, we will make it 80 x 20. To position it over the 60, 60 central spot, we move it to 20, 50:

```
bar.Resize(fyne.NewSize(80, 20))
bar.Move(fyne.NewPos(20, 50))
```

With this code in place, we end the function by returning the container we have made:

```
return c
```

This completes the definition of our sign. To display it, we call the makeSign() function and pass it to SetContent(). We can also turn off the default padding on the window, as our content is not reaching the edge of our container. Because we are using a manual layout, we also need to resize the window to show the items we have positioned:

```
w.SetContent(makeSign())
w.SetPadded(false)
w.Resize(fyne.NewSize(120, 120))
```

With this code in place, you can run the application in the usual way, but this time we shall force the dark theme so that our white border stands out (we will look at themes in more detail in *Chapter 5, Widget Library and Themes*):

```
Chapter03/canvas$ FYNE_THEME=dark go run main.go
```

You should see the following window with the sign crisp against a darker-colored background:

Figure 3.5 – Our road sign, created from a circle and rectangle

> **Note: manual layouts do not automatically resize**
>
> When using a manual layout, as explored in this example, it will not resize when the window changes size. This functionality is provided by using a layout algorithm, which we will explore in *Chapter 4, Layout and File Handling*.

In this section, we explored the canvas elements and how to draw them. They created crisp, clean output, but we did not explore how that works. In the next section, we look at what scalable rendering is and how it creates such high-quality output.

Scalable drawing primitives

As you probably realized from the previous example, all the items that we have rendered so far are vector graphics. This means that they are described by lines, curves, and high-level parameters instead of a collection of pixels. Because of this, these components are called **scalable** (like in **scalable vector graphics** (**SVG**) files), meaning that they can be drawn at any scale. The Fyne toolkit is a scalable toolkit, which means that a Fyne application can be drawn at any scale and render at a high quality.

Let's look at the text component in more detail, for example. We define a simple text component as before:

```
w.SetContent(canvas.NewText("Text", color.Black))
```

We can then place that line of code into the standard `main()` function that we wrote in the first section of this chapter, *Anatomy of a Fyne application*, and then run it. The output will be as expected—drawing text at the normal size—but if we override the preferred scale using `FYNE_SCALE`, we can see how the application would look if the user wanted much larger text:

Figure 3.6 – canvas.Text rendered with FYNE_SCALE=1 (left) and 3.5 (right)

Scaling a Fyne application in this way will not just change the font size, but will scale up every element of the user interface. This includes all standard graphics, widgets, and layouts. The standard themes also provide a set of icons (which we will explore more in *Chapter 5*, *Widget Library and Themes*), which are also scalable. We can see this by using a theme resource and the icon widget type:

```
w.SetContent(widget.NewIcon(theme.ContentCopyIcon()))
```

By adding the preceding line to the same `main` function, we can see how icons will scale to match the text demonstrated in the previous figure:

Figure 3.7 – widget.Icon rendered with FYNE_SCALE=1 (left) and 3.5 (right)

The size and position of elements will be scaled according to the canvas scale. We can now see how this coordinate system works.

Coordinate system

As you saw in the *Combining elements* section earlier, it is sometimes necessary to position or resize elements within a user interface. Doing so for a scalable output can be a little complicated, as we are not measuring content using pixels. And so, Fyne uses a device-independent coordinate system that Android developers may be familiar with.

A 1 x 1 size in Fyne (written as `fyne.NewSize(1, 1)`) may represent more than (or fewer than) 1 output pixel. If the scale was 3 (as it is for many modern smart phones), then the 1 x 1 square will likely use nine output pixels. As the toolkit is designed for scalable output, the result will not be a *pixelated* output, like it could be with older graphical toolkits that simply multiply the size of each pixel. The rendered output will continue to look crisp, as we saw in *Figure 3.6* and *Figure 3.7*.

A fully scalable user interface has huge benefits when working with display devices of varying pixel densities, and it allows users to choose a preferred zoom level for all application components. However, sometimes we need to work with nonscalable elements, such as bitmap (pixel based) images, or our application may need to use every pixel available for high-definition graphical output. We will explore this in the next section.

Pixel output – rendering images

For the reasons outlined in the previous section, it is recommended that you use scalable graphics (normally SVG files) for icons and other image-based components of user interfaces. It will sometimes be necessary to work with bitmap graphics (those defined by a collection of pixels rather than graphical features). If you are loading and managing images, or if you want to display detailed graphical elements using every pixel available, then this section contains important information about how to proceed.

Images

Image content in Fyne defines graphical content that will normally stretch or shrink to the space that is allocated to it. Loading a bitmap image (which has dimensions that are defined by the number of pixels) into a scalable output may not provide the expected outcome. The rendered output of a size defined in pixels will vary depending on the scale of the output device or user preferences. For this reason, Fyne does not normally set a minimum size for images that are being loaded.

Images can be loaded from files, resources, data streams, and more. Each file should have a unique name (or path), which makes it possible for performance improvements through caching. Loading an image from the filesystem should be done through the Fyne `storage` API to avoid any platform-specific code or assumptions (this is explored in detail in *Chapter 4, Layout and File Handling*). You can use `storage.NewFileURI` to get the reference to a file path if that is needed. For example, to load an image from a file path, you would call the following:

```
img := canvas.NewImageFromURI(storage.NewFileURI(path))
```

To define how a loaded image should be displayed in your app, you can set the `ImageFill` field in your `canvas.Image` object. It will be one of the following values:

- `canvas.ImageFillStretch`: The default value. This will adjust the image dimensions up or down to match the image object size, adjusting the **aspect ratio** (the ratio of width to height values), which can cause images to look squashed or stretched.

- `canvas.ImageFillContain`: This fill option will retain the aspect ratio so that the image is not distorted and draws it as the largest size possible that can fit within the image size. This will usually leave a space on two of the edges so that the image is centered in the available space.

- `canvas.ImageFillOriginal`: In original fill mode, the image will display using one output pixel for each pixel in the image. Although this seems desirable, it is important to note that it's visible size will vary based on the device because of the variation of pixel density. Using this value will also ensure that sufficient space will be reserved to draw the required number of pixels. If it is likely that the image will be larger than the space available, be sure to wrap it in a scroll container (discussed in *Chapter 5, Widget Library and Themes*).

As mentioned in the fill modes, the output size of an image cannot be known by looking at the image file, and so your application will probably need to specify how large the image should be. Normally, this will be controlled by a layout—images would be expanded according to the type of layout. Another way to do this is to call `SetMinSize()` to ensure that the image never gets smaller than a specified (pixel independent) value. If you have used `ImageFillOriginal`, then this step will have been completed automatically.

If small images are used, but they occupy a large space, they may appear *pixelated*, depending on how far they have been scaled. It is recommended that you use images that contain sufficient pixels/detail so that they will be scaled down (shown smaller) when displayed instead of scaling them up; however, if you want the output to look pixelated (or retro), you can specify that a pixel-based scaling should be used to enhance this look:

```
Image.ScaleMode = canvas.ImageScalePixels
```

Note that the pixelated output described here does not apply if the image file was .svg. When a scalable image file is loaded, it will always be redrawn to the size requested, ensuring a high-quality output each time.

Images are not the only way to draw bitmap content; we can also include more dynamically created pixel content using the Raster type.

Raster

In some cases, an application may wish to display content using every available pixel so that high detail can be displayed, such as when showing wave forms or a 3D render. In these situations, we use the Raster widget. This is designed to show pixel-based output that is calculated dynamically instead of being loaded from a file.

A raster output will dynamically determine the content to display based on the number of pixels available in the space it occupies. Each time the space resizes, the widget will be asked to redraw itself. These requests for content are handled by generator functions.

In this example, we will explore how to display a checkerboard pattern:

1. Firstly, we declare a generator function—this will take the x, y parameter of the pixel requested and the width and height parameter of the overall area (in pixels), and return a color value, as shown here:

```
func generate(x, y, w, h int) color.Color {
    ...
}
```

2. We then want to determine which color our pixel is. The following calculation will return white if the x, y coordinate is in a 20 x 20 pixel square at the top left or any odd-numbered square along the row, and then the opposite squares on the following row. For these pixels, we specify a white color:

```
if (x/20)%2 == (y/20)%2 {
    return color.White
}
```

3. And for any other pixel, we will return black instead:

```
return color.Black
```

4. With the generator function defined, we can create a new `raster` widget that will use it to color the output pixels:

```
w.SetContent(canvas.NewRasterWithPixels(generate))
```

5. By reusing the same application launch code in the previous examples, we can load the app and display its window:

```
Chapter03/raster$ go run main.go
```

6. This will display the result shown in the following figures:

Figure 3.8 - canvas.Raster rendered with FYNE_SCALE=1 (left) and 3 . 5 (right)

Whether you change the scale or resize the window, you will see that the pattern repeats at the same size, always using squares that are 20 pixels wide and high. Before we complete this section, we should also look at how gradients are handled on the canvas.

Gradient

As we saw in *Figure 3.3*, the Fyne canvas is also capable of displaying gradients. Much like the raster in the previous section, a gradient will display using all the available pixels for the best output possible for each device. Adding a gradient, however, is much simpler than managing raster content.

There are two types of gradient: linear and radial.

Linear gradient

A `LinearGradient` displays an even transition from one color to another and is normally presented as horizontal or vertical. A vertical gradient changes color from the start at the top of the area and the end color at the bottom; each row of pixels will have the same color, creating a gradient area that transitions from top to bottom. A horizontal gradient performs the same operation, but with the start color at the left of the area and the end at the right, which means that each column of pixels will have the same color.

For example, the following lines would create a horizontal vertical gradient respectively from white to black using the provided convenience constructors:

```
canvas.NewHorizontalGradient(color.White, color.Black)
canvas.NewVerticalGradient(color.White, color.Black)
```

By passing each of these to `Window.SetContent`, as we have done with other examples, you can see the following result, with a horizontal gradient on the left and a vertical gradient on the right:

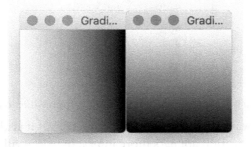

Figure 3.9 – Horizontal and vertical gradients

It is also possible to specify the exact angle of a linear gradient. The `NewLinearGradient` constructor takes a third parameter, the angle in degrees to orient. The vertical gradient is at `0` degrees and the horizontal is `270` degrees (equivalent to a `90`-degree counter-clockwise rotation). So, the usage of the horizontal gradient helper could also have been written as follows:

```
canvas.NewLinearGradient(color.White, color.Black, 270)
```

Sometimes, however, a gradient that forms a curve is required; for this, we use a radial gradient.

Radial gradient

A radial gradient is one where the start color is at the center of the area (although this can be offset using `CenterOffsetX` and `CenterOffsetY`) and progresses to the end color at the edge of the area. The gradient is drawn such that the end color fully appears within the bounds on the horizontal and vertical lines from the center of the gradient. This means that the corners of the area this gradient occupies will be outside of the gradient calculation; for this reason, it can be useful for the end color to be `color.Transparent`. We set up a gradient similar to the `LinearGradient` example from white to black, as follows:

```
canvas.NewRadialGradient(color.White, color.Black)
```

This code will result in the following image when placed in the contents of a window:

Figure 3.10 – A radial gradient transitioning from white to black

We have seen the various ways that we can output content, but it is also possible to animate it so that your application appears more interactive. We will see how to do this in the next section.

Animation of the draw properties

The Fyne `canvas` package also includes the facility to handle the animation of objects and properties. Using these APIs will help you to manage smooth transitions for a better user experience.

Animating a transition

An animation in Fyne, at its very basic level, is a function that will be called for every graphical frame that it runs. Once it is started, it will run as long as the `Duration` field specifies. A basic animation can be created using the following:

```
anim := fyne.NewAnimation(time.Duration, func(float32))
```

The animation that is returned from this constructor function can be started by calling `anim.Start()`. When the animation is started, its end time will be calculated based on the duration of time that passes. The callback that is passed in will be executed each time the graphics are updated. The `float32` parameter to this function will be `0.0` when it starts and `1.0` immediately before it ends; each intermediate call will have a value between these two.

To provide a more concrete illustration, we can set up a position animation. This is one of the helpful animations provided by the `canvas` package. It, like many others, takes two additional parameters: the `start` and `end` value of the animation. In this case, it expects a `start` and `end` `fyne.Position`. Note that that the `callback` function will provide the current position value instead of a `float32` *offset* parameter. We create a new position animation that will run for one second:

```
start := fyne.NewPos(10, 10)
end := fyne.NewPos(90, 10)
anim := canvas.NewPositionAnimation(start, end,
    time.Second, callback)
```

The `callback` function is responsible for applying the position value to a graphical object. In this case, we will create a text object that will move across the window:

```
text := canvas.NewText("Hi", color.Black)
callback := func(p fyne.Position) {
    text.Move(p)
    canvas.Refresh(text)
}
```

We then simply start this animation using the `Start()` method:

```
anim.Start()
```

These animations will run just once, but they can also be asked to loop.

Looping animations

Any animation can be set to repeat—this means that after the time duration lapses, it will start again at the beginning. To request this behavior, set the `RepeatCount` field on the `Animation` struct to `fyne.AnimationRepeatForever`:

```
anim.RepeatCount = fyne.AnimationRepeatForever
```

Setting `RepeatCount` to any number above `0` will specify how many times this animation should repeat.

```
anim.Start()
```

After a repeating animation is started, it will run until it is manually stopped (using `Animation.Stop()`) or the number of repeats specified in `RepeatCount` is reached.

There are many more animation APIs that can be used to control graphics and transitions. You can find more by looking for `NewXxxAnimation()` constructor functions in the `canvas` package.

Now that we have explored the graphical capabilities of the Fyne toolkit, we will put it together in a small game application.

Implementing a simple game

In the first example application of this book, we will see how the canvas elements come together by building the graphical elements of a *snake game* (for a history of this game, see the Wikipedia entry at `https://en.wikipedia.org/wiki/Snake_(video_game_genre)`. The main element of this game is a snake character that the user will control as it moves around the screen. We will build the snake from a row of rectangles and add animation elements to bring it to life. Let's start by drawing the initial screen.

Drawing a snake on screen

To start the work of displaying the game canvas, we will see create a simple snake that consists of a row of 10 green squares. Let's begin:

1. Firstly, we will create a setup function that will build the game screen. We will call this function `setupGame` and create an empty list that we will populate. The return from this method is a container with no layout so that we can later use a manual layout for the visual elements:

    ```
    func setupGame() *fyne.Container {
        var segments []fyne.CanvasObject
    ```

```
   . . .

       return container.NewWithoutLayout(segments...)
}
```

2. To set up the graphical elements, we will iterate through a loop of 10 elements (i from 0 to 9) and make a new Rectangle for each position. The elements are all created 10 x 10 in size and placed one above the other using the Move function. We will add them all to the segment slice created previously. This completes our setup code:

```
for i := 0; i < 10; i++ {
    r := canvas.NewRectangle(&color.RGBA{G: 0x66,
        A: 0xff})
    r.Resize(fyne.NewSize(10, 10))
    r.Move(fyne.NewPos(90, float32(50+i*10)))
    segments = append(segments, r)
}
```

The preceding color specification is using a hexadecimal format, where 0xff is the maximum for a channel and a missing channel (like red and blue in this code) defaults to 0. The result is a green of medium brightness.

3. With the graphical setup code, we can wrap this in the usual application load code, this time passing the result of setupGame() to the SetContent function. As this game will not have dynamic sizing, we will call SetFixedSize(true) so that the window cannot be resized:

```
func main() {
    a := app.New()

    w := a.NewWindow("Snake")
    w.SetContent(setupGame())
    w.Resize(fyne.NewSize(200, 200))
    w.SetFixedSize(true)
    w.ShowAndRun()
}
```

4. Now we can build and run the code in the usual way:

```
Chapter03/example$ go run main.go
```

5. You will see the following result:

Figure 3.11 – A simple snake drawn in the window

Next, we will bring the snake to life with some simple movement code.

Adding a timer to move the snake

The next step is to add some motion to the game. We will start with a simple timer that repositions the snake on screen:

1. To help manage the game state, we will define a new type to store the x, y value of each snake segment, named snakePart. We then make a slice that contains all of the elements, and this is what we will update as the snake moves around the screen. We will also define the game variable that we will use to refresh the screen when needed:

```
type snakePart struct {
    x, y float32
}

var (
    snakeParts  []snakePart
    game        *fyne.Container
)
```

2. Inside `setupGame`, we need to create the representation of snake segments, one for each of the rectangles we created before. Adding the following lines to the loop will set up the state:

```
seg := snakePart{9, float32(5 + i)}
snakeParts = append(snakeParts, seg)
```

3. To make sure that the game refreshes each time we move the snake, we need to move the rectangles and call `Refresh()`. We create a new function that will update the rectangles that we created earlier based on updated snake section information. We call this `refreshGame()`:

```
func refreshGame() {
    for i, seg := range snakeParts {
        rect := game.Objects[i]
        rect.Move(fyne.NewPos(seg.x*10, seg.y*10))
    }

    game.Refresh()
}
```

4. To run the main game loop, we need one more function that will use a timer to move the snake. We call this function `runGame`. This code waits 250 milliseconds and then moves the snake forward. To move it, we copy the position of each element from that of the element that is one segment further forward, working from the tail to the head. Lastly, the code moves the head to a new position, in this case further up the screen (by using `snakeParts[0].y--`). Refer to the following function:

```
func runGame() {
    for {
        time.Sleep(time.Millisecond * 250)
        for i := len(snakeParts) - 1; i >= 1; i-- {
            snakeParts[i] = snakeParts[i-1]
        }
        snakeParts[0].y--
        refreshGame()
    }
}
```

5. To start the game timer, we need to update the `main()` function. It must assign the `game` variable so that we can refresh it later, and it will start a new goroutine executing the `runGame` code. We do this by changing the `SetContent` and `ShowAndRun` calls to read as follows:

```
game = setupGame()
w.SetContent(game)

go runGame()
w.ShowAndRun()
```

6. Running the updated code will initially show the same screen, but the green shape will then move up the screen until it leaves the window:

```
Chapter03/example$ go run main.go
```

With the draw and basic movement code in place, we want to be able to control the game, which we will look at next.

Using keys to control direction

To control the snake's direction, we will need to handle some keyboard events. Unfortunately, this will be specific to desktop or mobile devices with a hardware keyboard; to add touchscreen controls would require using widgets (such as a button) that we will not explore until *Chapter 5, Widget Library and Themes*:

1. To start with, we define a new type (`moveType`) that will be used to describe the next direction in which to move. We use the Go built-in instruction `iota`, which is similar to `enum` in other languages. The `move` variable is then defined to hold the next move direction:

```
type moveType int

const (
    moveUp moveType = iota
    moveDown
    moveLeft
    moveRight
)

var move = moveUp
```

2. Next, we will translate from key events to the movement type that we just defined.
    Create a new `keyTyped` function as follows that will perform the keyboard
    mapping:

```
func keyTyped(e *fyne.KeyEvent) {
    switch e.Name {
    case fyne.KeyUp:
        move = moveUp
    case fyne.KeyDown:
        move = moveDown
    case fyne.KeyLeft:
        move = moveLeft
    case fyne.KeyRight:
        move = moveRight
    }
}
```

3. For the key events to be triggered, we must specify that this key handler should be
    used for the current window. We do this using the `SetOnKeyTyped()` function
    on the window's canvas:

```
w.Canvas().SetOnTypedKey(keyTyped)
```

4. To make the snake move according to these events, we need to update the
    `runGame()` function to apply the correct movement. In place of the line
    `snakeParts[0].y--` (just before `refreshGame()`), we add the following code
    that will position the head for each new move:

```
switch move {
case moveUp:
    snakeParts[0].y--
case moveDown:
    snakeParts[0].y++
case moveLeft:
    snakeParts[0].x--
case moveRight:
    snakeParts[0].x++
}
refreshGame()
```

5. We can now run the updated code sample to test the keyboard handling:

```
Chapter03/example$ go run main.go
```

6. After the app loads and you press the left and then down keys, you should see
    something like the following:

Figure 3.12 – The snake can move around the screen

Although this is now technically an animated game, we can make it smoother. Using the
animation API will allow us to draw smoother movements.

Animating the movement

The motion created by our run loop in the previous section provides the motion that
the game is based on, but it is not very smooth. In this last section, we will improve the
motion by using the animation API. We will create a new rectangle for the head segment
that will move ahead of the snake animation and also move the tail to its new position
smoothly. The rest of the elements can remain fixed. Let's see how this is done:

1. We first define a new rectangle that represents the moving head segment:

```
var head *canvas.Rectangle
```

2. We set this up by adding the following code to the setupGame function:

```
head = canvas.NewRectangle(&color.RGBA{G: 0x66, A:
    0xff})
head.Resize(fyne.NewSize(10, 10))
```

```
head.Move(fyne.NewPos(snakeParts[0].x*10,
    snakeParts[0].y*10))
segments = append(segments, head)
```

3. To start drawing the head in front of the current body position, we add the
    following code to the top of the `runGame` function so that the next segment is
    calculated in the movement before the snake reaches that position:

```
nextPart := snakePart{snakeParts[0].x,
    snakeParts[0].y - 1}
```

4. We set up our animations inside the `for` loop in the `runGame` function, before the
    timer pause. Firstly, we calculate where the head is and then its new position and set
    up a new animation to make that transition:

```
oldPos := fyne.NewPos(snakeParts[0].x*10,
    snakeParts[0].y*10)
newPos := fyne.NewPos(nextPart.x*10, next
    Part.y*10)
canvas.NewPositionAnimation(oldPos, newPos,
    time.Millisecond*250, func(p fyne.Position) {
    head.Move(p)
    canvas.Refresh(head)
}).Start()
```

5. We also create another animation to transition the tail to its new position, as
    follows:

```
end := len(snakeParts) - 1
canvas.NewPositionAnimation(
    fyne.NewPos(snakeParts[end].x*10,
        snakeParts[end].y*10),
    fyne.NewPos(snakeParts[end-1].x*10,
        snakeParts[end-1].y*10),
    time.Millisecond*250,
    func(p fyne.Position) {
        tail := game.Objects[end]
        tail.Move(p)
        canvas.Refresh(tail)
    }).Start()
```

6. After the `time.Sleep` in our game loop, we need to use the new `nextPart` variable to set up the new head position, as follows:

    ```
    snakeParts[0] = nextPart
    refreshGame()
    ```

7. After the refresh line, we need to update the movement calculations to set up `nextPart` ready for the next movement:

    ```
    switch move {
    case moveUp:
        nextPart = snakePart{nextPart.x,
            nextPart.y - 1}
    case moveDown:
        nextPart = snakePart{nextPart.x,
            nextPart.y + 1}
    case moveLeft:
        nextPart = snakePart{nextPart.x - 1,
            nextPart.y}
    case moveRight:
        nextPart = snakePart{nextPart.x + 1,
            nextPart.y}
    }
    ```

8. Run the updated code and you will see this behavior, but with a smooth transition along the path:

    ```
    Chapter03/example$ go run main.go
    ```

For the complete code for this example, you can use the book's GitHub repository at `https://github.com/PacktPublishing/Building-Cross-Platform-GUI-Applications-with-Fyne/tree/master/Chapter03/example`.

Although many more features could be added to this example, we have explored all the application and canvas operations required to build a full game.

Summary

In this chapter, we started our journey with the Fyne toolkit, exploring how it is organized and how a Fyne application operates. We saw how it uses vector graphics to create a high-quality output at any resolution, allowing it to scale well over desktop computers, smart phones, and more.

We explored the features of the canvas package and saw how it can be used to draw individual elements to the screen. By combining these graphical primitives using `fyne.Container`, we were able to draw more complex output to our window. We also saw how the animation APIs can be used to display transitions of an object's size, position, and other properties.

To bring this knowledge together, we built a small snake game that displayed elements to the screen and animated them based on user input. Although we could add many more features and graphical polish to this game, we will move on to other topics.

In the next chapter, we will explore how layout algorithms can manage the contents of a window and the best practices of file handling that are needed to build an image browser application.

4
Layout and
File Handling

In the previous chapter, we learned how the main drawing aspects of the Fyne toolkit are organized and how an application can work directly with `CanvasObject` items on a window canvas. This was sufficient information to set up a small game, but once applications add the presentation of lots of information or require user input and workflows, they typically require more complex user interface designs. In this chapter, we look at how an application user interface is structured, covering the following:

- Arranging a `Container` item using built-in layout algorithms
- Creating custom layout algorithms
- Handling files in a way that works across all platforms, desktop, and mobile

With this knowledge, we will build an application for browsing photographs. Let's get started!

Technical requirements

This chapter has the same requirements as *Chapter 3*, *Windows, Canvas, and Drawing*, which is to have the Fyne toolkit installed. For more information, please refer to the previous chapter.

The full source code for this chapter can be found at `https://github.com/PacktPublishing/Building-Cross-Platform-GUI-Applications-with-Fyne/tree/master/Chapter04`.

Laying out containers

As we saw in the previous chapter, a Fyne canvas is made up of `CanvasObject`, `Container`, and `Widget` items (although `Container` and `Widget` items are both `CanvasObject` items as well!). To be able to display multiple elements, we must use the `Container` type, which groups a number of `CanvasObject` items (which can also be `Widget` items or additional `Container` items). To manage the size and position of each item inside a container, we use an implementation of `Layout`, which is passed to the container at creation using the `container.New(layout, items)` constructor function.

There are many ways that an application may want to lay out its components and in this section we will explore the different ways that can be achieved. Layouts are not always required, however, and so first we will look at when you might not need to use a layout and how to handle size and placement manually instead.

Manual layout

Before we explore layout algorithms, it is possible to manage a container without the use of a layout—this is called **manual layout** and is done using `container.NewWithoutLayout(items)`.

When using a container without a layout, the developer must position and size all elements within the container manually, using the `Move()` and `Resize()` methods. In this mode, the developer is responsible for adapting the positions and sizes to the current size of the container.

Let's take a look at the following code:

```
square := canvas.NewRectangle(color.Black)
circle := canvas.NewCircle(color.Transparent)
circle.StrokeColor = &color.Gray{128}
circle.StrokeWidth = 5
box := container.NewWithoutLayout()(square, circle)
```

```
square.Move(fyne.NewPos(10, 10))
square.Resize(fyne.NewSize(90, 90))
circle.Move(fyne.NewPos(70, 70))
circle.Resize(fyne.NewSize(120, 120))
box.Resize(fyne.NewSize(200, 200))
```

The code we just saw sets up a `Rectangle` item and a `Circle` item inside a container, resizes them to be around half their size, and then positions them to have a small amount of overlap. You can see from the following figure that the elements are drawn in the order that they are passed to the container:

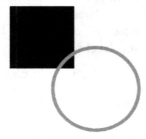

Figure 4.1 – Manual layout in a container

Once set, these sizes and positions will not change unless we add more code to modify their positions.

> **Important note**
>
> Note that there is no resized event published, so if you want to automatically adjust when the container is resized, you should consider building a custom layout, described under *Providing a custom layout* later in this chapter.

Using a layout manager

In essence, a layout manager is the same as the manual move and resize code we just saw, with the difference that it operates on a list of `CanvasObject` items (those that are the children of `Container`). A layout manager has two responsibilities:

- To control the size and position of each element
- To determine the minimum size that the container should accept

When a `Container` item is resized, the layout that it is using will be asked to reposition all the child components. A layout algorithm may choose to scale elements according to the new size or to reposition the elements. Alternatively, it may decide to flow elements or adapt the layout according to whether the available space is taller or wider. In this way, setting a layout on a container can provide a responsive user interface based on screen size or even device orientation.

When we lay out interface components, it is common to wish to separate elements by inserting some clear space. This is called *padding* and in Fyne, you can find the standard padding size using the `theme.Padding()` function. You can find more about the `theme` package in *Chapter 5*, *Widget Library and Themes*. The standard layouts listed in the next section all include the standard padding between elements. Note that typically, a container layout will not use padding on the outside edge as that will be provided by the parent container, or the window canvas for top-level containers.

Containers using a layout can be created using the `container.New` function:

```
container.New(layout, items)
```

When using a container with a layout, it is not usually required to call `Resize` as we did before because it will be initially sized to at least the minimum size.

Hidden objects

One additional consideration when selecting a layout, or writing layout code yourself, is that objects may not always be visible. A `CanvasObject` item may be hidden for two reasons:

- A developer called `Hide()` on that object.
- It is inside a `Container` item that has similarly had `Hide()` invoked.

Typically, a layout algorithm will skip hidden elements when calculating the minimum size or laying out elements. Each of the standard layouts we will see next will skip hidden elements rather than leaving empty space where those items would otherwise have appeared.

We have seen how layouts work in arranging the components of an application. To make building complex user interfaces as simple as possible, there are standard layouts available that cover most common user interface arrangements.

Standard layouts

As there are many standard layout algorithms, the Fyne toolkit includes a collection of standard implementations in the `layout` package. By importing this package, you can apply these layouts to any `Container` in your application:

```
import "fyne.io/fyne/v2/layout"
```

Each of the layouts is examined in detail in this section. Although a container can only have a single layout, there is no limit to the number of containers you can have nested inside each other, and so we look at combining different layouts at the end of this section.

MaxLayout

MaxLayout (or **maximum layout**) is the simplest of all the built-in layout algorithms. Its purpose is to ensure that all child elements of a container take up the full space of that container:

Figure 4.2 – MaxLayout in a container

This is most commonly used to align one element over another, such as a text item over a background color rectangle. When using this layout, it is important to list the container elements in the correct order; each will be drawn over the other, and so the last item in the list will be drawn on top:

```
myContainer := container.New(layout.NewMaxLayout(), …)
```

CenterLayout

CenterLayout can be helpful when an item of a specified minimum size should be centered within the available space, both horizontally and vertically:

Figure 4.3 – CenterLayout adding space around items

As with `MaxLayout`, each element within the container will be drawn on top of the previous one, but the size will be set as the minimum for each of the elements instead of filling the available space:

```
myContainer := container.New(layout.NewCenterLayout(), …)
```

PaddedLayout

PaddedLayout helps when you wish to inset content by the theme-defined padding value. The content element will be centered in the container by the standard padding on all sides, as shown in the following figure:

Figure 4.4 – PaddedLayout adding a small space around items

As with `MaxLayout`, each element within the container will be drawn on top of the previous one, all with the same size, but in this case slightly smaller than the container:

```
myContainer := container.New(layout.NewPaddedLayout(), …)
```

BoxLayout

The box layout has two varieties, `HBoxLayout` (horizontal—for arranging items in a row) and `VBoxLayout` (vertical—for arranging items in a list). Each of the box layouts follows a similar algorithm: it creates a linear flow of elements where they are packed (horizontally or vertically) while maintaining a consistent height or width.

Items listed in a horizontal box will have the width set to each item's minimum, but will share the same height, which is the maximum of all the elements' minimum height:

Figure 4.5 – HBoxLayout aligning three elements in a row

Items in a vertical box all have the same width (the largest of all the minimum widths), while shrinking to each element's minimum height:

Figure 4.6 – VBoxLayout stacking elements in a column

This approach allows items of differing sizes to appear uniform without wasting any space in the container. The syntax for each of these is as follows:

```
myContainer := container.New(layout.NewHBoxLayout(), …)
myContainer := container.New(layout.NewVBoxLayout(), …)
```

FormLayout

FormLayout is used by the form widget, but it can be useful on its own when you wish to label items in a container. There should be an even number of elements added; the first of each pair will be on the left, being as narrow as the component allows. The remaining horizontal space will be taken up by the second of each pair:

Figure 4.7 – FormLayout pairing items for labeling

Here's an example of using `FormLayout` (assuming an even number of parameters to be added):

```
myContainer := container.New(layout.NewFormLayout(), …)
```

GridLayout

The basic **GridLayout** is designed to divide a container into as many equal spaces as the number of child elements in the container.

For a container with two columns and three child items, a second row will be created but not completely filled:

Figure 4.8 – Three elements in a two-column GridLayout

When creating a grid layout, the developer will specify the number of columns or rows to use, and the items will be arranged accordingly. At the end of each row or column, the layout will wrap and create a new one. The number of rows (or columns) will depend upon the number of elements. For example, let's take the following illustration:

```
myContainer := container.New(layout.
NewGridLayoutWithColumns(2), …)
myContainer := container.New(layout.NewGridLayoutWithRows(2),
…)
```

The grid layout has an additional mode that can help to adapt to different output devices. It is common on mobile devices to show items in a single column when held in portrait or a single row in landscape orientation. To enable this, use `NewAdaptiveGridLayout`; the parameter to this constructor represents the number of rows you wish to have in vertical arrangement or columns when horizontal. This layout will rearrange its `Container` when a mobile device is rotated, as seen here:

```
myContainer := container.New(layout.NewAdaptiveGridLayout(3),
…)
```

GridWrapLayout

Another variant of using a grid is when you would like elements to automatically flow to new rows as a container is resized (for example, a file manager or list of image thumbnails). For this scenario, Fyne provides a grid wrap layout. In a wrapped grid in which every child element will be resized to the specified size, they will then be arranged in a row until the next item does not fit, at which point a new row will be created for further elements.

For example, here is a grid wrap container that is wider than three items of the specified size:

Figure 4.9 – Fixed elements in GridWrapLayout

> **GridWrapLayout and MinSize**
>
> It is important to note that this layout, unlike all the others, will not check each item's `MinSize`. The developer should thus be careful to ensure that it is large enough, or that elements included will truncate their elements (such as text) to avoid overflow.

Here is an example using a grid wrap layout:

```
myContainer := container.New(layout.NewGridWrapLayout(fyne.
    NewSize(120, 120), …)
```

BorderLayout

The most commonly used layout in arranging an application is probably **BorderLayout**. This layout algorithm will arrange specified elements at the top, bottom, left, and right edges of a container. The top and bottom items will be resized to their minimum height but stretched horizontally, and items on the left and right will be squashed to their minimum width and expanded vertically. Any elements in the container that are not specified as belonging to one of the edges will be sized to fill the available space inside the borders. This is commonly used to position toolbars at the top, footers at the bottom, and file lists on the left. Any edges you wish to leave blank should have nil instead:

Figure 4.10 – BorderLayout with top and left areas set

BorderLayout parameters

Note that for `BorderLayout`, some elements must be specified twice—the layout parameters specify where an element should be positioned, but the list of items to the container control what will be visible. If you find an item not appearing, be sure that it is specified in both places.

The following code shows how to set up a border container with `header` at the top and `files` positioned to the left of `content`:

```
myContainer := container.New(layout.NewBorderLayout(header,
    nil, files, nil), header, files, content)
```

Combining layouts

To build more complicated application structures, it will be necessary to use multiple layouts within your user interface. As each container has a single layout, we achieve this by nesting different containers. This can be done as many times as required. For example, take a look at the following figure:

Figure 4.11 – Multiple containers with different layouts

For the previous illustration, we have used a container with VBoxLayout for the left panel, HBoxLayout for the top, and GridWrapLayout for the central container, all inside BorderLayout, as follows:

```
top := container.New(layout.NewHBoxLayout(), ...)
left := container.New(layout.NewVBoxLayout(), ...)
content := container.New(layout.NewGridWrapLayout(fyne.
    NewSize(40, 40)), ...)
combined := container.New(layout.NewBorderLayout(top, nil,
    left, nil), top, left, content)
```

Using the container package

All of the preceding examples use a built-in Layout type to configure the contents of fyne.Container. To help manage more complex layout configurations (we will see more in *Chapter 5, Widget Library and Themes*), there are many helpful constructor functions in the container package. For example, instead of container.New(layout.NewBorderLayout(...)...) we could use container. NewBorder(...), which can lead to clearer code.

Providing a custom layout

If the standard layouts, or a combination of them, do not accommodate the needs of your user interface, it is possible to build a custom layout and pass that into a container instead.

Any type that implements the fyne.Layout interface can be used as a Container layout. This interface has just two methods that need to be implemented, as shown here:

```
// Layout defines how CanvasObjects may be laid out in a
// specified Size.
type Layout interface {
    // Layout will manipulate the listed CanvasObjects Size
    // and Position to fit within the specified size.
    Layout([]CanvasObject, Size)
    // MinSize calculates the smallest size that will fit the
    // listed
    // CanvasObjects using this Layout algorithm.
    MinSize(objects []CanvasObject) Size
}
```

As you can see, this interface codifies the earlier description that a layout manager will need to determine the minimum size of a container as well as handling the positioning of each element in a container. As the contents of a container can change from time to time, it is possible that the slice of CanvasObject elements passed to Layout or MinSize may change. Hence, a custom layout should avoid caching references to individual elements. In certain situations (such as BorderLayout, which we saw earlier), the layout may explicitly hold an object reference. If your layout works in this way, it is important to remember that the item may not exist within the slice of objects to lay out.

Most layouts should also skip hidden elements when calculating the minimum size or layout. There are some exceptions, however, particularly if elements are likely to be shown and hidden regularly. For example, a tab-based layout that shows only one content element at a time could cause windows to expand if hidden elements are larger than the ones that are currently visible. In this situation, it will be good for the user if layouts consider hidden elements in the MinSize code, even if they are not positioned in Layout.

We shall create a short example of writing a custom layout:

1. This type, named diagonal, will position items of a container in a diagonal line from the top left down to the bottom right. We first implement MinSize() to return the sum of all visible objects (so that there is space to display them all in a diagonal line):

```go
type diagonal struct{}

func (d *diagonal) MinSize(items []fyne.CanvasObject)
fyne.Size {
    total := fyne.NewSize(0, 0)
    for _, obj := range items {
        if !obj.Visible() {
            continue
        }

        total = total.Add(obj.MinSize())
    }
    return total
}
```

2. We then add the `Layout()` method, which is responsible for actually positioning each object. In this version, we simply declare a `topLeft` variable and position each visible object, adding to the value each time we have positioned and sized an element:

```
func (d *diagonal) Layout(items []fyne.CanvasObject, size
fyne.Size) {
    topLeft := fyne.NewPos(0, 0)
    for _, obj := range items {
        if !obj.Visible() {
            continue
        }

        size := obj.MinSize()

        obj.Move(topLeft)
        obj.Resize(size)
        topLeft = topLeft.Add(fyne.NewPos(size.Width,
            size.Height))
    }
}
```

3. To apply this layout to a container, you would simply use the following:

```
item1 := canvas.NewRectangle(color.Black)
item1.SetMinSize(fyne.NewSize(35, 35))
item2 := canvas.NewRectangle(&color.Gray{128})
item2.SetMinSize(fyne.NewSize(35, 35))
item3 := canvas.NewRectangle(color.Black)
item3.SetMinSize(fyne.NewSize(35, 35))
myContainer := container.New(&diagonal{}, item1, item2,
    item3)
```

Here's what we get:

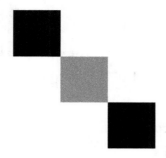

Figure 4.12 – Simple diagonal layout

The example we saw sets up a static layout. The minimum size set for each item sets up the minimum size for `Container`, which no longer expands. An improved version of this layout would calculate the amount of extra space (the difference between the container's `MinSize()` and the `size` argument passed into the `Layout()` function). An updated version of the `Layout()` function looks like this:

```
func (d *diagonal) Layout(items []fyne.CanvasObject, size fyne.
Size) {
    topLeft := fyne.NewPos(0, 0)

    visibleCount := 0
    for _, obj := range items {
        if !obj.Visible() {
            continue
        }

        visibleCount++
    }

    min := d.MinSize(items)
    extraX := (size.Width - min.Width)/visibleCount
    extraY := (size.Height - min.Height)/visibleCount

    for _, obj := range items {
        if !obj.Visible() {
            continue
        }

        size := obj.MinSize()
```

```
                    size = size.Add(fyne.NewSize(extraX, extraY))

                    obj.Move(topLeft)
                    obj.Resize(size)
                    topLeft = topLeft.Add(fyne.NewPos(size.Width,
                        size.Height))
                }
        }
```

Here's what we get after running the code:

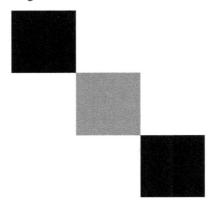

Figure 4.13 – Diagonal layout expanding to fill space

With this more advanced code, we no longer need to control the container with a minimum size for all items (though items would normally have a minimum size). In fact, we could just resize the container (or the app window) instead, as follows:

```
item1 := canvas.NewRectangle(color.Black)
item2 := canvas.NewRectangle(&color.Gray{128})
item3 := canvas.NewRectangle(color.Black)
myContainer := canvas.New(&diagonal{}, item1, item2, item3)
myContainer.Resize(fyne.NewSize(120, 120))
```

Now that we have explored the basics of how to lay out an application, we can start to look at bringing this together in a real application. The example that we will explore is an image browsing app that will lay out images and their metadata. However, before we can do this, we need to learn about file handling in a cross-platform context. If app developers assume users will have a filesystem or structure that matches their development system, it may not work on other devices, so understanding how to do this well is essential for ensuring that apps work well across all devices.

Cross-platform file handling

The Go standard library has excellent support for file handling across its supported platforms. The os package allows access to the filesystem (files and directories) and utility packages such as filepath that help to parse and manage locations using the current operating system's semantics. While these operations are likely useful on most devices, they do not extend as well to non-desktop devices where a traditional filesystem is not what the end user is presented with.

Consider mobile devices, for example. Both iOS and Android have a traditional filesystem internally, but the filesystem is not completely available to the device user, nor is it the only source of file data. An application will typically only have access to its own sandbox directory—reading and writing files outside of this space is not permitted—and on iOS, you may even need to request special permissions before accessing it. In addition to that, users now expect to be able to open data from other sources. For example, a file-sharing application such as Dropbox could provide a source of files that a user may wish to pass into your application, but this data is not accessible using standard file handling.

For these reasons, the Fyne toolkit includes a simple storage abstraction that allows your application to handle data from any source, while managing permissions and security considerations for you. This interaction uses the concept of a **URI** to replace traditional file paths, allowing apps to operate without direct access to files and directories.

URI

At the core of the file handling abstraction is fyne.URI (here **URI** stands for **Uniform Resource Identifier**). A URI will be familiar to most computer users as it looks exactly like a web URL, with the small difference that it does not always start with http:// or https://. A URI may represent a filesystem object (where it would start with file://), a data stream from another app (where it may begin content://), or a remote resource (such as sftp:// for a Secure File Transfer Protocol connection).

Like the os.File type, fyne.URI is a reference to a resource, though it does not keep that resource open, so it may be passed around your application without issues. The underlying string representation of this URI can be accessed using the String() method. Use this if you wish to store the URI reference for later use, for example, in a configuration file or database. If you have a URI string representation, the original URI object can be accessed using utilities in the storage package, as follows:

```
uriString := "file:///home/user/file.txt"
myUri, err := storage.ParseURI(uriString)
```

Reading and writing

Accessing files when you are not certain where they are stored is a little more complicated than the traditional `os.Open()`; however, the Fyne `storage` package provides functionality to handle this. The two main functions for data access are `OpenFileFromURI` and `SaveFileToURI`, as shown in this excerpt from the package:

```go
// OpenFileFromURI loads a file read stream from a resource
// identifier.
func OpenFileFromURI(uri fyne.URI) (fyne.URIReadCloser, error)
{
    return fyne.CurrentApp().Driver().FileReaderForURI(uri)
}

// SaveFileToURI loads a file write stream to a resource
// identifier.
func SaveFileToURI(uri fyne.URI) (fyne.URIWriteCloser, error) {
    return fyne.CurrentApp().Driver().FileWriterForURI(uri)
}
```

Each of these functions take a URI (as described in the preceding code) for the location and returns `URIReadCloser` or `URIWriteCloser` on success and `error` if the operation failed.

As suggested by their names, these return types implement `io.ReadCloser` and `io.WriteCloser` with the addition of a `URI()` function to query the original resource identifier. You may not recognize these `io` interfaces, but you will have used them through `os.File`. This similarity means that you can use `URIReadCloser` in many places where you would have passed a file for a read operation, or `URIWriteCloser` if you were writing data.

If you are processing the read or write operations yourself, it is important to remember to call `Close()` upon completion (as with any `io.Closer` stream). This is most commonly ensured by calling `defer reader.Close()` after checking for any error. The following code shows a simple example of reading a file from a URI:

```go
uri := storage.NewFileURI("/home/user/file.txt")
read, err := storage.OpenFileFromURI(uri)
if err != nil {
    log.Println("Unable to open file \""+uri.
        String()+"\"", err)
    return
}

defer read.Close()
data, err := ioutil.ReadAll(read)
```

```
if err != nil {
    log.Println("Unable to read text", err)
    return
}

log.Println("Loaded data:", string(data))
```

User file selection

The most common way for an application to open a file, at least initially, would be to prompt the user to choose the file they wish to open. The standard file-open dialog is available to provide this feature. An application can call dialog.ShowFileOpen, which will ask the user to select a file (with optional file filters). The chosen file will be returned through a callback function as URIReadCloser, as described earlier. If you wish to store a reference to the chosen file, you can use the URI() method to return the identifier. The following code shows this in action:

```
dialog.ShowFileOpen(func(reader fyne.URIReadCloser, err error) {
    if err != nil { // there was an error - tell user
        dialog.ShowError(err, win)
        return
    }
    if reader == nil { // user cancelled
        return
    }


    // we have a URIReadCloser - handle reading the file
    // (remember to call Close())
    fileOpened(reader)
}, win)
```

Similarly, there is dialog.ShowFileSave to start a file-write workflow such as the common *Save As* feature. For more information on the dialog package, see the *Dialogs* section in *Chapter 5, Widget Library and Themes*.

ListableURI

In some applications, it may be necessary to open a resource that contains other resources (just like a directory of files). For these situations, there is another type, `fyne.ListableURI`, which provides a `List()` method that returns a slice of `URI` items. This can be used in combination with `dialog.ShowDirectoryOpen`, which will return the user's selected location as `ListableURI`.

Let's take a look at an example:

```go
dialog.ShowFolderOpen(func(dir fyne.ListableURI, err error) {
    if err != nil { // there was an error - tell user
        dialog.ShowError(err, win)
        return
    }

    if dir == nil { // user cancelled
        return
    }

    log.Println("Listing dir", dir.Name())
    for _, item := range dir.List() {
        log.Println("Item name", item.Name())
    }
}, win)
```

As you can see in this example, once the user has made their selection, `ListableURI` is passed to our code. We can then iterate through the URI of each item inside the directory or collection using `range List()`. If you already have the name of a directory, then you can use `storage.ListerForURI(storage.NewFileURI(dirPath))`.

Let's put layouts and file handling into action. We will now build a simple image browsing application using all that we've seen till now.

Implementing an image browser application

This application will load a directory that contains some images, provide a summary of the content in a status bar at the bottom of the window, and use most of the space to show each image. The images will be loaded as thumbnails (smaller versions of the images) and we will display the image information under each thumbnail.

Creating the layout

To start this example, we will create the layout of the application and the image items that will display in the central grid. Let's understand each of these actions in detail:

1. First, we set up the image items. We wish to have the image name underneath the image. While this could be positioned manually, the items will be more responsive to changes in size if we use `BorderLayout`. We will create a `canvas.Text` element in the `bottom` position and use `canvas.Rectangle` to represent the image that we will load later:

```
func makeImageItem() fyne.CanvasObject {
    label := canvas.NewText("label", color.Gray{128})
    label.Alignment = fyne.TextAlignCenter
    img := canvas.NewRectangle(color.Black)
    return container.NewBorder(nil, label, nil, nil,
        img)
}
```

2. For the main application, we need to create the grid to contain image thumbnails as well as the status panel, which will be positioned later on. For the image grid, we will use `GridWrapLayout`. This version of a grid layout sizes all elements to a specified size and as the available space increases, the number of visible items will also increase. In this case, the user could increase the window size to see more images.

3. As we have not yet loaded the directory, we will fake the number of items (hardcoded to three by iterating over {1, 2, 3}). We create a list of items, calling `makeImageItem` for each one. This list is then passed to `NewGridWrap` after the `size` parameter (which is the size used for each item—a behavior specific to the grid wrap layout):

```
func makeImageGrid() fyne.CanvasObject {
    items := []fyne.CanvasObject{}
    for range []int{1, 2, 3} {
        img := makeImageItem()
        items = append(items, img)
    }

    cellSize := fyne.NewSize(160, 120)
    return container.NewGridWrap(cellSize, items...)
}
```

4. To start, we will just create a text placeholder for the status for the purpose of laying out the app:

```
func makeStatus() fyne.CanvasObject {
    return canvas.NewText("status", color.Gray{128})
}
```

5. And finally, we will create a new container once again using `BorderLayout` to arrange the status bar beneath the rest of the content. By placing the image grid in the central space of `BorderLayout`, it will fill any available space as desired:

```
func makeUI() fyne.CanvasObject {
    status := makeStatus()
    content := makeImageGrid()
    return container.NewBorder(nil, status, nil, nil,
        content)
}
```

6. To complete the application, we just need a short `main()` function that loads the Fyne application and creates a window, and we will resize it to larger than the minimum size so that the image grid layout will expand to multiple columns:

```
func main() {
    a := app.New()
    w := a.NewWindow("Image Browser")

    w.SetContent(makeUI())
    w.Resize(fyne.NewSize(480, 360))
    w.ShowAndRun()
}
```

7. All we have to do now is run the combined code:

```
Chapter04/example$ go run main.go
```

8. Running this will show the following window, ready for some real data and images to be loaded:

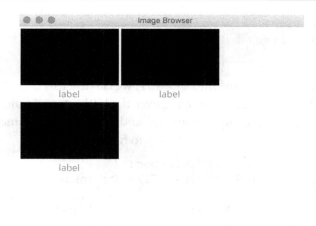

Figure 4.14 – The basic layout of our photos app

Listing a directory

Before we can load images, we need to establish which directory we are loading when the application starts. Let's take a look at the steps to do just that:

1. From the main() function, we will call a new startDirectory (that parses the app flags or falls back to the current working directory) and pass this into the makeUI() function. The directory path is converted to ListableURI by calling ListerForURI and NewFileURI:

```
func startDirectory() fyne.ListableURI {
    flag.Parse()
    if len(flag.Args()) < 1 {
        cwd, _ := os.Getwd()
        list, _ := storage.ListerForURI(
            storage.NewFileURI(cwd))
        return list
    }

    dir, err := filepath.Abs(flag.Arg(0))
    if errr != nil {
        log.Println("Could not find directory", dir
        cwd, _ := os.Getwd
        list, _ := storage.ListerForURI(
            storage.NewFileURI(cwd))
        return list
    }
    list, _ := storage.ListerForURI(storage.
```

```
NewFileURI(dir))
    return list
}
```

2. Once `ListableURI` is passed into `makeUI`, we can use this `dir.List()`
    and filter it for image files before ranging over the URIs. A new function,
    `filterImages`, will take the directory list and return a slice of image URIs. To do
    this, a small `isImage()` function will help to filter:

```
func isImage(file fyne.URI) bool {
    ext := strings.ToLower(file.Extension())

    return ext == ".png" || ext == ".jpg" ||
        ext == ".jpeg" || ext == ".gif"
}

func filterImages(files []fyne.URI) []fyne.URI {
    images := []fyne.URI{}

    for _, file := range files {
        if isImage(file) {
            images = append(images, file)
        }
    }

    return images
}
```

3. With a slice of `fyne.URI` representing the images, we can update the status and
    image grid creation functions as well as updating the image label to use `URI.`
    `Name()` under each image placeholder:

```
func makeImageGrid(images []fyne.URI) fyne.CanvasObject {
    items := []fyne.CanvasObject{}

    for range images {
        img := makeImageItem()
        items = append(items, img)
    }

    cellSize := fyne.NewSize(160, 120)
    return container.NewGridWrap(cellSize, items...)
}

func makeStatus(dir fyne.ListableURI, images []fyne.URI)
```

```
fyne.CanvasObject {
    status := fmt.Sprintf("Directory %s, %d items",
        dir.Name(), len(images))
    return canvas.NewText(status, color.Gray{128})
}

func makeUI(dir fyne.ListableURI) fyne.CanvasObject {
    list, err := dir.List()
    if err != nil {
        log.Println("Error listing directory", err)
    }
    images := filterImages(list)
    status := makeStatus(dir, images)
    content := makeImageGrid(images)
    return container.NewBorder(
        (nil, status, nil, nil, content)
}
```

Loading the images

Let's now look at the steps to load images into our application:

1. To start with, we create a simple image load method that accepts a URI and returns
    *canvas.Image. The new loadImage function will then be used instead of the
    placeholder rectangle:

```
func loadImage(u fyne.URI) fyne.CanvasObject {
    read, err := storage.OpenFileFromURI(u)
    if err != nil {
        log.Println("Error opening image", err)
        return canvas.NewRectangle(color.Black)
    }
    res, err :=
        storage.LoadResourceFromURI(read.URI())
    if err != nil {
        log.Println("Error reading image", err)
        return canvas.NewRectangle(color.Black)
    }

    img := canvas.NewImageFromResource(res)
    img.FillMode = canvas.ImageFillContain
    return img
}
```

2. The `makeImage` function should be updated to pass the `URI` item as follows:

```
func makeImageItem(u fyne.URI) fyne.CanvasObject {
```

3. Then the line that creates the image inside the `makeImageItem` function as a rectangle can be replaced with the image created:

```
img := loadImage(u)
```

4. In the `loadImage` function, before returning `canvas.Image`, we changed `FillMode` from the default (`canvas.ImageFillStretch`) to `canvas.ImageFillContain` so that the image aspect ratio will be respected and the images will fit within the available space:

Figure 4.15 – Images and names loaded into the layout

This code works as expected, as we can see in the figure, but it can be slow. We are loading the images before continuing with the user interface load. This does not make for a good user experience, so let's improve this situation by using background image loading.

Loading the app faster

To avoid image sizes slowing down the loading of our user interface, we need to complete the construction of the application UI before the images load. This is called asynchronous (or background) loading and can be powerful if your app needs to use large amounts of resources.

The easiest way to load all the images in the background would be to start many goroutines. But, when displaying a large directory, that could become very slow indeed. Instead, we will use a single image load goroutine that will process one image at a time. (As an exercise, if you are feeling adventurous, you could expand this to process eight or more images at a time.)

Let's now take a look at how to do this:

1. To track the image loads, we will create a new type called `bgImageLoad` that will reference the URI of the image to load and the `*canvas.Image` item that it should be loaded into. We additionally need to create a channel (we'll name it `loads`) that will enqueue the items to load. We buffer this at `1024` items, which represents a large directory—an implementation to handle unbounded numbers of files might need us to be a little smarter:

    ```
    type bgImageLoad struct {
        uri fyne.URI
        img *canvas.Image
    }

    var loads = make(chan bgImageLoad, 1024)
    ```

2. When loading images in this updated version, we will create an empty Fyne `*canvas.Image` that will later have the image loaded. We then queue the details of this image URI, for loading once the goroutine is able to process it:

    ```
    func loadImage(u fyne.URI) fyne.CanvasObject {
        img := canvas.NewImageFromResource(nil)
        img.FillMode = canvas.ImageFillContain

        loads <- bgImageLoad{u, img}
        return img
    }
    ```

3. We move the image load code to a new `doLoadImage` function that will run in the background. In this version, we want to do all of the slow parts of image loading; so, we load and decode the image, convert it to a Go in-memory image to display, and leave the user interface much more responsive to updates, resizing, and so on.

    The new function, `doLoadImages`, will range over all the items being added to the channel and call `doLoadImage` to load them one at a time. The image load code will refresh the image `CanvasObject` after loading the raw data, so each item appears as it is loaded:

    ```
    func doLoadImage(u fyne.URI, img *canvas.Image) {
        read, err := storage.OpenFileFromURI(u)
        if err != nil {
            log.Println("Error opening image", err)
            return
        }
    ```

```
        defer read.Close()
        raw, _, err := image.Decode(read)
        if err != nil {
            log.Println("Error decoding image", err)
            return
        }

        img.Image = scaleImage(raw)
        img.Refresh()
    }

func doLoadImages() {
    for load := range loads {
        doLoadImage(load.uri, load.img)
    }
}
```

4. To make sure that the images are loaded, we launch doLoadImages as a goroutine within the main() function:

```
func main() {
...
    go doLoadImages()
    w.ShowAndRun()
}
```

5. Finally, in the preceding code, we referenced scaleImage. This means that each image we display is a smaller version of the full-sized image. This is necessary when the directory we browse contains very large images. The toolkit will attempt to paint very large images quite small, which can be very slow. Instead, we reduce the size of our images to fit inside the space available in each grid cell. We used larger numbers (twice the cell size) so that high-density displays still give a good-looking result.

6. The following code snippet makes use of the helpful github.com/nfnt/ resize package to scale images. Although the image package in Go is often helpful, it does not contain efficient scaling routines. We use this library and request Lanczos3 interpolation, which provides a balance between speed and quality when downscaling images:

```
func scaleImage(img image.Image) image.Image {
    return resize.Thumbnail(320, 240, img,
        resize.Lanczos3)
}
```

The `resize.Thumbnail` function creates a smaller image that fits within the stated size, which is ideal for our purpose so we can avoid worrying about aspect ratios and calculations.

Using the updated code will create a quick-to-load and responsive user interface for even large directories containing very large images. There we have it: resizing, which could have been slow when using full-sized images, is now much faster!

Creating a custom layout for the image elements

The space taken up by labels in this example could be a bit wasted, so let's make a custom layout that writes the text over the bottom edge of each image. We will use a semi-transparent background to make the text more readable and a small gradient to blend from the text background to the image.

To build a custom layout, we need to define a type (`itemLayout` in this case) that implements the `MinSize` and `Layout` functions from the `fyne.Layout` interface. As the background, gradient, and text all have special positions, we will save a reference to these canvas objects so that they can be arranged later:

```go
type itemLayout struct {
    bg, text, gradient fyne.CanvasObject
}

func (i *itemLayout) MinSize(_ []fyne.CanvasObject) fyne.Size {
    return fyne.NewSize(160, 120)
}

func (i *itemLayout) Layout(objs []fyne.CanvasObject, size
fyne.Size) {
    textHeight := float32(22)
    for _, o := range objs {
        if o == i.text {
            o.Move(fyne.NewPos(0, size.Height-textHeight))
            o.Resize(fyne.NewSize(size.Width, textHeight))
        } else if o == i.bg {
            o.Move(fyne.NewPos(0, size.Height-textHeight))
            o.Resize(fyne.NewSize(size.Width, textHeight))
        } else if o == i.gradient {
            o.Move(fyne.NewPos(0, size.Height-
                (textHeight*1.5)))
            o.Resize(fyne.NewSize(size.Width, textHeight/2))
        } else {
            o.Move(fyne.NewPos(0, 0))
            o.Resize(size)
```

```
            }
        }
    }
```

This code will ensure that each of the elements of our container is positioned in the correct place. `text` and `bg` are bottom aligned with `gradient` positioned above the text background. Any other element (in this case, our image thumbnail) will be positioned at the fill size that the layout is asked to fill.

To use this layout, we update the `makeImageItem` function to use `&itemLayout` as the container layout. Into this constructor, we pass a new `canvas.Rectangle` and `canvas.Gradient` to be used for the effect described previously. It is important to pass the image before the text background and pass the text element last to `NewContainerWithLayout`, as this sets up the order that these elements will be drawn:

```
func makeImageItem(u fyne.URI) fyne.CanvasObject {
    label := canvas.NewText(u.Name(), color.Gray{128})
    label.Alignment = fyne.TextAlignCenter

    bgColor := &color.NRGBA{R: 255, G: 255, B: 255, A: 224}
    bg := canvas.NewRectangle(bgColor)
    fade := canvas.NewLinearGradient(color.Transparent,
        bgColor, 0)
    return container.New(
        &itemLayout{text: label, bg: bg, gradient: fade},
        loadImage(u), bg, fade, label)
}
```

With these changes, we can run the code once more and see how our new layout makes each image preview larger in the same amount of space, while giving the application a little flair:

Figure 4.16 – Custom layout for images and their labels

Last of all, you may have noticed that directories with many images will force the window to expand, and so you may want to add scrolling to the grid container. To do so, we will use one of the helpers from the container package mentioned earlier, adding `container.Scroll` around the image grid container. It just requires replacing the `content` creation line of the `makeUI` function with this:

```
content := container.NewScroll(makeImageGrid(images))
```

Changing a directory

In addition to loading a specified directory, we may wish to allow users to open a different directory once the application is open. To add this functionality, we will use the `SetMainMenu` function on `Window`, which sets up a structure to populate a menu bar.

Using the `NewMainMenu`, `NewMenu`, and `NewMenuItem` helper functions in the `fyne` package, we set up a structure that defines the **File | Open Directory...** menu and will call `chooseDirectory` when clicked (we also pass in the current window so we can show a dialog from that function). The following code gets added to the `main()` function just before `Window.ShowAndRun()`:

```
w.SetMainMenu(fyne.NewMainMenu(fyne.NewMenu("File",
    fyne.NewMenuItem("Open Directory...", func() {
        chooseDirectory(w)
    })))))
```

To support this menu operation, we need to create the `chooseDirectory` function. This will call `dialog.ShowDirectoryOpen`, which asks the user to select a directory on their computer. This functions much like the `ShowFileOpen` call we explored before, with the exception that the parameter returned in the callback is `ListableURI` instead of `URIReadCloser`. Using this parameter (after checking for any error), we can call `makeUI` with this new location and update our whole application's user interface:

```
func chooseDirectory(w fyne.Window) {
    dialog.ShowFolderOpen(func(dir fyne.ListableURI, err
    error) {
        if err != nil {
            dialog.ShowError(err, w)
            return
        }
        w.SetContent(makeUI(dir)) // this re-loads our
                                  // application
    }, w)
}
```

If we were building a more complex application, then simply calling `Window.SetContent` would not be the most efficient approach. In that situation, we would save a reference to the main `fyne.Container` and update just the image grid instead of the entire application. However, the final version of our application should look like the following screenshot:

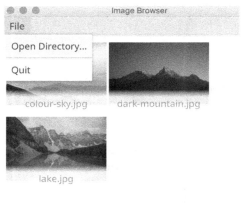

Figure 4.17 – Adding a main menu

Note that when running on macOS, the default behavior is to show menus in the desktop menu bar—this can be overridden using the `no_native_menus` build tag, as follows:

```
$ go run -tags no_native_menus main.go Images/Desktop
```

Summary

This chapter stepped through the details of how layouts work, the details of all the built-in layouts in the toolkit, and when to use them. We also saw how simple it is to combine multiple layouts and created our own custom layout to add a bit of flair to our image browsing application.

We also explored how to adapt file handling code to work across all platforms using the `URI` and `ListableURI` types. Using this knowledge, our image browsing application is now compatible with all desktop and mobile platforms. With this knowledge of how to lay out applications and avoid assumptions about a traditional filesystem, you can now ensure that your apps will function correctly on any supported platforms, mobile, desktop, and beyond.

While we have created a complete application using just canvas primitives and layouts, it is possible to build much more complex applications using the `widget` package, which we will look at in the next chapter.

5
Widget Library and Themes

A large part of the Fyne toolkit is its library of standard widgets, which provides simple visual elements, manages user input, and handles application workflows. These widgets handle how information and user input is displayed, as well as container options for organizing the user interface and managing standard workflows. The themes that come with the Fyne toolkit support both light and dark versions, both of which support user color preferences while adapting all the user interface elements so that they look great in both modes.

In this chapter, we're going to explore the widgets available in the Fyne toolkit and how to use them. We will be covering the following topics:

- Exploring the design of the Fyne Widget API
- Introducing the basic widgets
- Grouping with the collection widgets
- Adding structure with container widgets
- Using common dialogs

By the end of this chapter, you will be familiar with all the Fyne widgets and how the theme capabilities work to control their appearance. By combining widgets and containers, you will have built your first complete Fyne-based graphical user interface.

Technical requirements

This chapter has the same requirements as *Chapter 3, Windows, Canvas, and Drawing* – that is, you must have the Fyne toolkit installed and a Go and C compiler working. For more information, please refer to the previous chapter.

The full source code for this chapter can be found at `https://github.com/ PacktPublishing/Building-Cross-Platform-GUI-Applications-with- Fyne/tree/master/Chapter05`.

Exploring the design of the Widget API

As we described in *Chapter 2, The Future According to Fyne*, its APIs are designed to convey semantic meaning rather than a list of features. This is followed on by the Widget definition, whereby we add APIs that describe behavior and hide the details of rendering. The interface that all widgets must implement is simply an extension of the basic `CanvasObject` object (introduced in *Chapter 3, Window, Canvas, and Drawing*), which adds a `CreateRenderer()` method. It is defined in the source code as follows:

```
// Widget defines the standard behaviors of any widget.
// This extends the CanvasObject - a widget behaves in
// the same basic way but will encapsulate many child
// objects to create the rendered widget.
type Widget interface {
        CanvasObject
        CreateRenderer() WidgetRenderer
}
```

The new `CreateRenderer()` method is used by Fyne to determine how the widget should look. There is no public API for accessing the current renderer – instead, state is set in a `Widget` and each renderer will refresh its output to match this state. This design strongly encourages APIs to focus on behavior and intent rather than directly manipulating the graphical output (which could quickly lead to inconsistent or unusable applications).

Focus on behavior

By enforcing a separation between widget state and its visual representation, each `Widget` is forced to expose an API that describes behavior or intent instead of visual attributes or internal details. This is important for continuing the design principle of a semantic API, which leads to a more concise API focused on expected outcomes over graphical tweaks.

Commonly, changes in state will need to be reflected by a `WidgetRender` updating in some way. How the graphical representation will change is controlled by the renderer, and it is triggered by calling `Widget.Refresh()`. This refresh is normally handled by the widget code; for example, the `Button.SetText()` function's code is as follows:

```
// SetText allows the button label to be changed
func (b *Button) SetText(text string) {
    b.Text = text

    b.Refresh()
}
```

The call to `Refresh()` queues a request asking for the widget rendering to be updated, which will result in graphical updates in the next update of the screen. On some widgets, calling the `Refresh` function can cause a lot of calculations. If you have lots of changes to apply to a widget, you may not wish it to refresh after every line. To help with this, there is also a direct field access approach that ensures it doesn't need to be called as often.

Methods versus field access

As discussed in the previous section, refreshing a widget, especially a complex one, can be time-consuming, and if many updates must be performed, the user may notice a slight delay. If a developer wishes to update multiple aspects of a widget, then it is possible (though perhaps unlikely) that one change may be applied before a visual redraw and the others happen just after. Although another draw will occur, it may be noticeable that they did not change together. Consider the following code:

```
func updateMyButton(b *widget.Button) {
    b.SetText("sometext")
    b.SetIcon(someResource)
}
```

When using method-based updates, the `SetText()` and `SetIcon()` calls will refresh the widget, possibly causing the slight delay mentioned previously. It is recommended to call `Refresh()` only when needed; to make this possible, a developer can access the widget state directly and refresh the object manually. This is known as **field-based access** as we directly change the exported fields of a widget. For example, we could rewrite the preceding code like so:

```
func updateMyButton(b *widget.Button) {
    b.Text = "sometext"
    b.Icon = someResource

    b.Refresh()
}
```

By taking this approach, we ensure that `Refresh()` is called just once so that both changed elements will redraw at the same time. This offers a smoother result for the user and will lead to lower CPU usage as well.

Rendering a widget

The `CreateRenderer()` method mentioned earlier will return a new renderer instance that defines how the widget will be presented on-screen. The toolkit is responsible for calling this method and it will cache the result while the widget is visible. Developers should not call this directly as the result will have no connection to what is displayed.

The exact lifetime of a renderer varies and is a combination of the widget's visibility, its parent's visibility, and whether the window is currently shown. The `WidgetRenderer` definition for a widget may be unloaded during its application life cycle. A new instance will be requested if the widget becomes visible again at a later date. If a renderer is no longer needed, then its `Destroy()` method will be called. The full definition of a widget renderer is as follows and has been taken from the API documentation:

```
// WidgetRenderer defines the behavior of a widget's
// implementation. This is returned from a widget's main
// object through the CreateRenderer() function.
type WidgetRenderer interface {
    Layout(Size)
    MinSize()  Size
    Refresh()
    Objects()  []CanvasObject
    Destroy()
}
```

The first two methods in the `WidgetRenderer` definition (`Layout()` and `MinSize()`) should be familiar from *Chapter 4*, *Layout and File Handling* – they define a container layout. In this instance, the container is this widget and the objects that are being controlled are the visual components used to render the widget – they are returned from the `Objects()` method. The `Refresh()` method of a `WidgetRenderer` is called internally when a visible widget needs to be redrawn.

The `Objects()` method call returns a list of each `CanvasObject` required to render the widget it describes. This is a slice of items from the `canvas` package such as `Text`, `Rectangle`, and `Line`. These items will be arranged using the layout methods described previously to create the final widget presentations. It is important that the color of each element match, or blend with, the current theme. When the `Refresh()` function is called, it may be in response to a theme change, so any custom values should be updated accordingly in that code.

Now that we know how widgets work, let's look at what is built into the toolkit.

Introducing the basic widgets

The most used package in Fyne (or any GUI toolkit) is likely the `widget` package. This contains all the standard widgets that will be useful to most graphical apps. The collection is split into basic widgets (for simple data display or user input) and collection widgets (`List`, `Table`, and `Tree`) that are used to display large or more complex data. In this section, we'll step through the basic widgets in alphabetical order to see how they look and how to add them to an app.

Accordion

The accordion widget is used to fit large amounts of content into a small area by showing and hiding items so that only one of the child elements is visible at a time. Each item has a header button that is used to show or hide the content below it. This can be seen in the following image, which shows an accordion widget in the light and the dark themes:

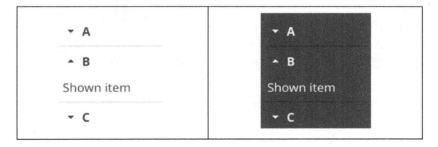

Figure 5.1 – An accordion widget with item B expanded, shown in the light and dark themes

To create an `Accordion` widget, you must pass a list of `AccordionItem` objects that specify the title and details for each element in the accordion. You can also optionally specify the `Open` value, which, if set to `true`, will show the content by default. The following code block was used to create the previous `Accordion` images:

```
acc := widget.NewAccordion(
    widget.NewAccordionItem("A", widget.NewLabel("Hidden")),
    widget.NewAccordionItem("B", widget.NewLabel("Shown
        item")),
    widget.NewAccordionItem("C", widget.NewLabel("End")),
)
acc.Items[1].Open = true
```

By default, an `Accordion` widget shows only one item at a time. To allow any number of items to be shown, you can set the `MultiOpen` field to `true`.

Button

The `Button` widget provides a standard press button that can be activated by clicking it with a mouse (or by using the tap gesture on a touch screen). A button can contain text or icon content, or both. The `Button` constructor function also takes an anonymous `func` that is executed when the button is tapped.

You can create a button using the `widget.NewButton` or `widget.NewButtonWithIcon` constructor functions. It is recommended to use the built-in theme icons where possible (for more information, please see the *Themes* section later in this chapter), as follows:

```
widget.NewButtonWithIcon("Cancel", theme.CancelIcon(), func()
{})
```

The button in each theme is shown in the following image:

Figure 5.2 – A button with its icon shown in light and dark themes

Most buttons have the same look, but if you need one to stand out, you can set it to have a high importance style by setting Button.Importance = widget.HighImportance, as shown in the following image:

Figure 5.3 – A high importance button in both light and dark themes

The color used to represent high importance widgets will vary based on the theme and can even be set by the user. Another possible value for Button.Importance is widget.LowImportance, which reduces the visual impact of the display, for example, by removing the shadow shown in the previous two images.

Card

When the elements of the user interface relate to each other, it can be useful to group them together. This can be helpful when a group of items should be titled or where many different data elements want a larger preview than a simple list. A collection of Card widgets may be added to a Container with a grid layout to arrange similar items that display different content (such as search results or a preview of media items).

A card widget can be created using the NewCard constructor function, which takes a title and subtitle string, as well as a CanvasObject content parameter. You can also specify the Image field after constructing or by creating the struct manually, as shown in the following code block:

```
widget.NewCard("Card Title", "Subtitle",
    widget.NewLabel("Content"))
&widget.Card(Title: "Card Title",
    Subtitle: "Subtitle",
    Image: canvas.NewImageFromResource(theme.FyneIcon()))
```

You can see the preceding code utilizing the light and dark themes in the following image:

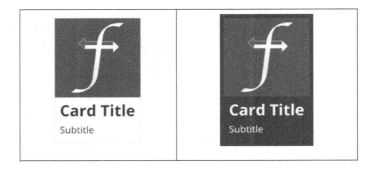

Figure 5.4 – The card widget showing its titles and an image in light and dark modes

All the fields that were used in the preceding code are optional. An empty string for either of the titles will remove it from the display, and a nil image or content removes it as well (you can see the missing content in the previous image).

Check

Checkbox functionality is provided by the Check widget. It has two states: checked and unchecked (the default). The constructor takes a callback function, func(bool), that will be called whenever the checked state changes, passing the current state (true means checked). The code for this is as follows:

```
widget.NewCheck("Check", func(bool) {})
```

You can also manually set the checked state by calling the SetChecked method and passing true (or false to change it back to unchecked):

```
check.SetChecked(true)
```

The different checked states, one in each theme, are shown in the following image:

Figure 5.5 – A light theme checkbox checked, and a dark theme checkbox unchecked

The checked indicator will adapt to the current theme, just like the text does.

Entry

Text input is primarily added through the `Entry` widget, which provides free text entry. A content hint can be added to a field by setting a `Placeholder`, and the text content can be set directly with the `Text` field or the `SetText ()` method. Entries can be created with one of the following constructors. The first is a normal text field, the second is a password entry, and the third is used for multiline input (defaults to 3 lines):

```
widget.NewEntry()
widget.NewPasswordEntry()
widget.NewMultilineEntry()
```

The following image shows the standard and password entry widgets using the default themes:

Figure 5.6 – Entry and PasswordEntry widgets in light and dark themes

The entry widget also supports validation, which can be used to give the user feedback on the current text. The different states can be seen in the following image:

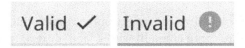

Figure 5.7 – Validation status of success and failure using the light theme

While typing, an entry with a validator set will provide positive feedback when the content is valid. Invalid content will show a warning when the user stops editing the input.

FileIcon

In addition to using static icons (see the Icon section, later in this section), the toolkit can display appropriate icons for different types of files. The `FileIcon` widget was created to make this common task much easier, by loading one of the standard icon resources and showing the file extension inside it. The icon matches the size of the standard `widget.Icon` and its image and text will be updated to reflect the specified URI. `FileIcon` widgets can be created by specifying the URI of a file resource, as follows:

```
file := storage.NewFileURI("images/myimage.png")
widget.NewFileIcon(file)
```

The preceding code will be rendered as follows, depending on the current theme:

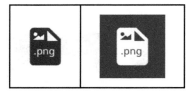

Figure 5.8 – The Fileicon widget showing the symbol and extension of a PNG image

`FileIcon` widgets will vary their appearance, setting icons appropriate for the file type.

Form

The form widget is useful for labeling and laying out various input elements. Each row of a form typically contains an input element with a label next to it. If the form has an `OnSubmit` or `OnCancel` function set, then it will automatically generate an additional row containing the appropriate action buttons.

You can create a form by passing a list of `FormItem` objects into the `NewForm()` constructor function, as follows:

```
form := widget.NewForm(
    widget.NewFormItem("Username", widget.NewEntry()),
    widget.NewFormItem("Password", widget.NewPasswordEntry()),
)
form.OnCancel = func() {
    fmt.Println("Cancelled")
}
form.OnSubmit = func() {
    fmt.Println("Form submitted")
}
```

You can also call `Form.Append()` to add elements later. The form will look as follows, depending on the current theme:

Figure 5.9 – A login form shown in light and dark themes

An additional benefit of grouping input elements into a form is that it can help ensure that only validated input is submitted. If any of the widgets have a validator set, then the form will only allow submission once all the validators have passed.

Hyperlink

In some situations, it can be helpful to provide a tappable URL string that opens a web page in the user's browser. For this use case, there is a `Hyperlink` widget that accepts a URL parameter for the page to open when it's tapped. To create a hyperlink widget, we may need to parse a URL, as shown in the following code snippet:

```
href, _ := url.Parse("https://fyne.io")
widget.NewHyperlink("fyne.io", href)
```

The widget looks as follows:

Figure 5.10 – A hyperlink widget using the light and dark themes

Icon

Although you can add a `canvas.Image` to a user interface directly, it can sometimes be useful to have a consistent size. The `Icon` widget provides this, loading a standard resource and setting its minimum size to the theme-defined icon size. The icon will also update to match the current theme. We can create an icon that shows the standard paste image using the following code:

```
widget.NewIcon(theme.ContentPasteIcon())
```

The preceding code will be rendered as follows, depending on the current theme:

Figure 5.11 – The icon widget showing the paste icon in the light and dark themes

Next, we move on to the label widget.

Label

Displaying text is normally recommended using the Label widget. It is possible to use canvas.Text but, as we saw in *Chapter 3, Windows, Canvas, and Drawing*, those elements use a developer-defined color – in most user interfaces, it is preferable for the text to match the current theme, which is what the Label widget takes care of:

```
widget.NewLabel("Text label")
```

The label widget in each theme looks as follows:

Figure 5.12 – A text label in the light and dark themes

This widget also supports additional text formatting such as newline, word wrap, and truncation. A string containing the newline character (\n) will be broken into a second line and be fully visible, whereas setting Label.Wrapping = fyne.TextWrapWord (or fyne.TextWrapBreak) will automatically add new lines when the widget's width requires it. Setting the value to fyne.TextWrapTruncate will simply remove overflowing text from the display.

Pop-up menu

The PopUpMenu widget is helpful for displaying context menus or allowing a user to choose from options where a built-in widget is not providing the desired functionality. To create a menu in this way, you need a fyne.Menu data structure that describes the options available (this is the same structure we used in *Chapter 3, Windows, Canvas, and Drawing*, for the main menu). In this case, the title of the menu can be empty as it will not be displayed.

Once the menu has been defined, the `ShowPopUpMenuAtPosition` utility function is the easiest way to display this to the user. This function takes an absolute position in the window and displays the menu in the top-left corner. As with other pop-up elements, this can be dismissed by tapping outside of the content shown, thereby returning the user to the previous state. You can use the following code to do this:

```
menu := fyne.NewMenu("", fyne.NewMenuItem("An item", func()
{}))
pos := fyne.NewPosition(20, 20)
widget.ShowPopUpMenuAtPosition(menu, myWindow.Canvas(), pos)
```

The preceding code will create a new menu over the current content of the window. The result will look as follows:

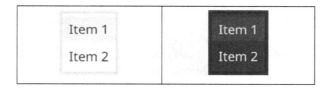

Figure 5.13 – PopUpMenu shown in the light and dark themes

It is also possible to use the `Menu` widget directly here. The `widget.NewMenu` constructor function can be used to render a menu without it creating an overlay like `PopUpMenu` does.

ProgressBar

If an application needs to indicate that a process will take some time to perform, then you can use the `ProgressBar` widget. There are two variations: `widget.ProgressBar` and `widget.ProgressBarInfinite`. A regular progress bar shows a current `Value` ranging from `Min` to `Max` (default *0* to *1*), and the developer is responsible for setting the value as the process progresses. When using the infinite progress bar, there is no intrinsic value, so the output renders an animation that indicates an activity (a change in the value) for an undefined duration. We can create a progress bar using one of the following lines:

```
bar1 := widget.NewProgressBar()
bar2 := widget.NewProgressBarInfinite()
```

These two versions of the progress bar look as follows, with `bar1` on the left and `bar2` on the right:

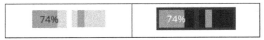

Figure 5.14 – The standard and infinite progress bars shown with light and dark themes

If you wish to add a different text overlay for the regular progress widget, then you can use the `TextFormatter` field, setting a function which returns a string value. This can be formatted based on the widget's state or something different such as the `loading...` string.

RadioGroup

The radio group widget is the most common way of requesting user input. The outcome is similar to the `Select` widget (discussed next), but the options are all visible. The `RadioGroup` widget can be created by specifying a slice of string values that will be listed as options. The second parameter is a callback that will execute each time the selection is changed, passing the new value into the function:

```
widget.NewRadioGroup([]string{"Item 1", "Item 2"}, func(s string) {
    fmt.Println("Selected", s)
})
```

The radio group widget looks as follows in the different themes:

Figure 5.15– A radio group with two options, top selected, in the light and dark themes

Select

Like the preceding `RadioGroup` widget, the `Select` widget allows the user to choose one item from a list. It is more common to use `Select` when the list of options is long or the space that's available is small. This widget appears as a button showing the current value. When tapped, this will show a popup menu that lists the options that are available:

```
widget.NewSelect([]string{"Item 1", "Item 2"}, func(s string) {
    fmt.Println("Selected", s)
})
```

The Select widget looks as follows in the different themes:

Figure 5.16 – The select widget shown in the light and dark themes

Next is the SelectEntry widget.

SelectEntry

SelectEntry is much like the Select widget described previously, except that it allows user-defined options as well. This can be done by presenting an Entry widget with a select-style drop-down icon that lists the specified options. Because the current value can change on every keystroke, the callback for this widget is configured like Entry rather than Select – it is not passed as a select change function in the constructor but can instead be set on the OnChanged field of the widget:

```
widget.NewSelectEntry([]string{"Item 1", "Item 2"})
```

The SelectEntry widget looks as follows:

Figure 5.17 – The SelectEntry widget before input is added in the light and dark themes

Next is the Slider widget.

Slider

The Slider widget can be used for inputting a number within a range, particularly when the exact number may not be known to the user. This can be useful, for example, when specifying brightness or volume – the number is not important to the end user, but it has a clear range from Min to Max.

A `Slider` widget can be created by specifying `Min` and `Max` values through the constructor. Its default value will be set to the minimum, and this can be changed by setting `Slider.Value`. It is also possible to specify a `Step` value, which defines the distance between each valid value. Without a defined step, any floating-point value between the minimum and maximum will be allowed. By specifying `Step`, you could, for example, accept only integer values. In this mode, the user may see the slider "jump" from one valid value to another as they slide the widget:

```
widget.NewSlider(0, 100)
```

The preceding code will create a simple slider widget set to the minimum value, as shown in the following image:

Figure 5.18 – The Slider widget shown at its minimum in the light and dark themes

TextGrid

Although the `Label` widget (described earlier) does provide some text formatting, there are some applications that require styles to be applied per-character. For syntax highlighting in a code editor or for showing error rows indicated on a console output, there is the `TextGrid` widget.

Inside a `TextGrid` widget, the content is split into each rune of the string representation, and each rune has a `TextGridStyle` applied to it. This style allows the foreground and background color to be specified for each character or cell of the grid. Additionally, each row of the grid can have a style specified. This style will be used for any cell that does not have its own style specified. If a cell has a character and a row style available, the two will be combined so that the foreground color that's been set on the cell will adopt a background color from the row; that is, unless the character style specifies both.

Despite the styles that allow specific colors to be set, there are a number of semantic style definitions that allow code to annotate intent instead of absolute colors. One of the most commonly used styles is `TextGridStyleWhitespace` that uses the theme definition to show characters in a muted color. Using the built-in styles, a developer can delegate to the current theme to define colors for each intent.

The `TextGrid` widget also provides common functionality for technical text displays, including `ShowLineNumbers`, which displays the line number at the beginning of each row. Also, `ShowWhitespace` can be set to true for a visual indicator of otherwise invisible spacing characters such as tab, space, and newline. The following code example illustrates some of the ways you can control text in a `TextGrid`:

```
grid := widget.NewTextGridFromString(
    "TextGrid\n  Content  ")
grid.SetStyleRange(0, 4, 0, 7,
    &widget.CustomTextGridStyle{BGColor:
        &color.NRGBA{R: 64, G: 64, B: 192, A: 128}})
grid.Rows[1].Style = &widget.CustomTextGridStyle{BGColor:
        &color.NRGBA{R: 64, G: 192, B: 64, A: 128}}

grid.ShowLineNumbers = true
grid.ShowWhitespace = true
```

The following image shows the result of the preceding code. Here, we can see that a background style has been applied to all the cells used in the 4 letters of *Grid* and that a row style has been applied to the second row (index 1). They also have line numbers and the whitespace options turned on:

Figure 5.19 – Styled content presented using TextGrid in the light and dark themes

The style for a cell can be assigned using the `SetStyle` method. However, when the style needs to be applied to many runes, developers can use the more efficient `SetStyleRange` utility method. The `SetRowStyle` method can assist in setting row styles, as shown in the previous image.

Toolbar

If there are lots of regularly accessed features in an application, the `Toolbar` widget can be an efficient way to present these options. The main elements of a toolbar are the `ToolbarAction` items, which are simple icons that, when tapped, execute a function parameter that's been passed to `NewToolbarAction`. To group action elements, you can use `ToolbarSeparator`, which creates a visual divider between the item on its left and right. Additionally, a gap can be created between actions using the `ToolbarSpacer` type. This will expand, causing the elements after it to be right aligned. Using one spacer will show items before it on the left and items after it on the right. Using two will mean that the elements between the spacers will be central in the toolbar.

To construct a toolbar containing four action elements and a separator, we can use the following code:

```
widget.NewToolbar(
    widget.NewToolbarAction(theme.MailComposeIcon(),
        func() {}),
    widget.NewToolbarSeparator(),
    widget.NewToolbarSpacer(),
    widget.NewToolbarAction(theme.ContentCutIcon(),
        func() {}),
    widget.NewToolbarAction(theme.ContentCopyIcon(),
        func() {}),
    widget.NewToolbarAction(theme.ContentPasteIcon(),
        func() {}),
)
```

The preceding code snippet results in the following. The following image shows it in both the light and dark themes:

Figure 5.20 – Toolbar widget with some possible icons in the light and dark themes

The widgets that we have explored so far are fairly standard and can be created with simple constructors, or through initializing the struct directly. Some of these take a callback function that can be used to inform us when an action has occurred.

In the next section, we'll look at some more complex widgets that are designed to manage thousands of sub-widgets. To do so, we will learn how they make use of more function parameters to query large datasets and efficiently display a subset of the data.

Grouping with the collection widgets

In this section, we will look at widgets that are designed to efficiently contain main widgets. Some of the widgets mentioned in the previous section do this, such as `Form` and `Toolbar`, but **collection widgets** support thousands of items (though they're not all visible at one time). These widgets are commonly used for displaying a huge numbers of options or navigating complex datasets.

Due to the requirement that collection widgets only show large amounts of data, they are designed to only show a small portion of the possible widget at a time. To do this, and to maintain great performance, they have a caching mechanism that makes their API a little more complex than the widgets we have seen previously.

Callbacks

Each of these widgets relies on a number of callback functions. The first of these functions will provide information on the dimensions of the data that the widget will display (for a more complete discussion on data, see *Chapter 6, Data Binding and Storage*). The second of these is responsible for creating visual elements that will be displayed later, while the third will load an item from the data into a previously created element.

Caching

The key to the performance of collection widgets is how they cache the graphical elements that repeat within them. The template objects referenced in collection widget constructors will be reused as the user scrolls the widget to maintain performance and keep up with the user actions.

The `List` widget (and other collection widgets) maintains an internal cache of recently used template elements that will have new data applied in preparation for the next rows becoming visible. It is the job of the application developer to optimize data retrieval so that any items that are close to those that are already visible will load quickly. We will see these concepts in use as we explore the various collection widgets that are available to us. First, we will look at the List widget.

List

The `List` widget is used to display a vertical list of items where each item has a similar look. The relevant data can be loaded once the widget has been created, so it can be helpful if the data is slow to load or complex to display. The widget will only load and display the elements that are visible, thereby displaying the elements of a large dataset quickly.

Callbacks

Each of the collection widgets uses callback functions to understand the data, load a template item, and update it when the data is loaded. Let's look at these in more detail:

- Understanding the data – the `Length` callback: The first callback function for `List` is the `length` callback, which returns the number of items in the data. This tells the widget how many rows it will need to manage. If more items are added to the dataset, this value can be updated, and next time the list refreshes, it will adjust accordingly.

- Loading a template – the `CreateItem` callback: The second callback function is used to generate a reusable graphical element that will load data. This is called a **template** item. The return value of this function is a `CanvasObject` and can be any type of `Widget`, `Container`, or item from the `canvas` package. The widget will call this function as many times as there are items visible on the screen. At this stage, they should contain just placeholder values. For example, in the images that follow, each row contains an icon and a label, so the returned template would probably be a container with a horizontal box layout, along with a default icon and placeholder text in the label. Although the user will never see the placeholder values, they are important as the size of a template configures the `List` component. The height of a template item will be used for the height of every row so that when it's multiplied by the result of the previous length, the callback will determine the overall scroll height of the list component. Additionally, the template width will specify the minimum width for the `List` component.

- Filling the template with data – the `UpdateItem` callback: Callback three is used to apply data to a template cell. It receives two parameters: the index of the data item to use and the `CanvasObject` template that we configured earlier. The purpose of this callback is to configure the item with the data that should be used at the specified index. The template that's used will be identical to the return object of the second parameter so that it can be cast appropriately.

Each collection widget has variations of the pattern described previously, as we will see in the *Table* and *Tree* sections later in this chapter.

Selection

One additional feature of the collection widgets is that they allow an element to be selected (by being tapped). In the list interface, the selected element is indicated by a marker at the leading edge, as shown in the preceding image. To be notified when an item is selected, you can set a func(ListItemID) callback on the OnSelected field, which will notify you of which item from the dataset was selected. The basic code for creating a list is as follows:

```
widget.NewList(
    func() int { return 3 },
    func() fyne.CanvasObject {
        icon := widget.NewIcon(theme.FileIcon())
        label := widget.NewLabel("List item x")
        return container.NewHBox(icon, label)
    },
    func(index ListItemID, template fyne.CanvasObject) {
        cont := template.(*fyne.Container)
        label := cont.Objects[1].(*widget.Label)
        label.SetText(fmt.Sprintf("List item %v", index))
    })
```

The code sample will generate the following output, once the second item has been tapped:

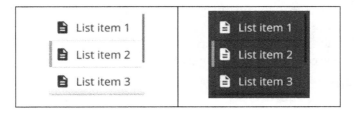

Figure 5.21 – List collection widget with an item selected in the light and dark themes

Table

The Table widget is a two-dimensional version of the List widget and is designed for showing large datasets with row and column aspects. It uses the same system for caching and callbacks as the List and Tree widgets.

Callbacks

The Table widget's callbacks are similar to those for List, but the data identifiers pass a row and column int to index the data structure. This means that the Length callback now returns (int, int).

The callback that sets up new graphical templates in the Table version (called CreateCell) takes no parameters and just returns a fyne.CanvasObject that will be cached for use in the display. This template is used to determine the default size of all cells, so make sure that it has a sensible minimum size. As with List, the element you return here will not be presented to the user but will be used for measurements and configuring the overall layout.

The last required callback is UpdateCell and is used to apply data to a template element. In the Table widget, this function passes a data identifier (TableCellID, which contains a Row and Col int) that indexes the data to apply, as well as the CanvasObject template. Developers should fill in the template with the appropriate data specified by the identifier. As with other collection widgets, it is recommended, where possible, to load related data so that when the user scrolls or expands an element, any data that takes a long time to load is ready to be displayed.

Selection

The Table widget supports a selected cell, as is indicated by a marker at the leading edge and header, as shown in the following image. To be notified of when an item has been selected, you can set the func (TableCellID) callback on the OnSelected field. This will notify you of which item from the dataset was selected by passing the identifying row and column. The basic code for creating a new table is as follows:

```
widget.NewTable(
    func() (int, int) { return 3, 3 },
    func() fyne.CanvasObject {
        return widget.NewLabel("Cell 0, 0")
    },
    func(id TableCellID, template fyne.CanvasObject) {
        label := template.(*widget.Label)
        label.SetText(fmt.Sprintf("Cell %d, %d", id.Row+1,
            id.Col+1))
    })
```

The preceding code will generate the following output, assuming cell 2, 1 is tapped to gain selection:

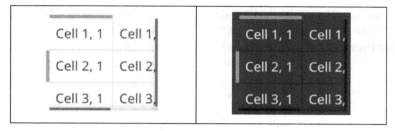

Figure 5.22 – Table collection widget showing selection in the light and dark themes

Tree

The Tree widget is very similar to the List widget, but with the added feature that each element can expand to show other items. This expansion is used to show a hierarchy, such as directories and files, categories and items, or other data with a parent-child relationship present.

Callbacks

The Tree widget's callbacks are similar to List and Table, but its more complex data structure means that the simple Length callback is replaced by ChildUIDs and IsBranch. The first of these callbacks will return a slice of TreeNodeID values (regular strings can be used) that contain the unique identifier for each item that exists under the specified node (passed in as a unique TreeNodeID). The second is called for each element while passing the unique ID. It should return true if it can container further nodes, or false otherwise.

The callback that sets up new graphical templates in the Tree version (called CreateNode) takes a bool parameter that represents if this is a branch (true, can expand) or a leaf (false, this is the end of the tree). This is useful if you want to use a different style for the branch and leaf elements within your tree. As with List, the element you return here will not be presented to the user but will be used for measurements and configuring the overall layout.

The last required callback is UpdateNode. This is used to apply data to a template element. In the Tree widget, this function passes the unique TreeNodeID identifier, a bool representing whether this is a branch or leaf template, and the CanvasObject template. Developers should fill in this template with the appropriate data specified by the identifier. As with other collection widgets, it is recommended, where possible, to load related data so that when the user scrolls or expands an element, data that loads slowly is ready to be displayed.

Additionally, regarding the callbacks that are required to manage content, the `Tree` widget allows developers to set the `OnBranchOpened` and `OnBranchClosed` callbacks so that they can track changes in the state of the tree. Both functions are of the `func(TreeNodeID)` type, where the parameter is the unique identifier of the data item.

Selection

The `Tree` widget also supports selected nodes. This is indicated by a marker at the leading edge, as shown in the following image. To be notified of when an item has been selected, you can set the `func(TreeNodeID)` callback on the `OnSelected` field. This will notify you of which item of the dataset was selected, while passing the unique identifier. The basic code required to show a tree is as follows. The first callback is returning the unique IDs of the children at each level:

```
func(uid TreeNodeID) []string {
    switch uid {
    case "":
        return []string{"cars", "trains"}
    case "cars":
        return []string{"ford", "tesla"}
    case "trains":
        return []string{"rocket", "tgv"}
    }
return ""
},
func(uid TreeNodeID) bool {
    return uid == "" || uid == "cars" || uid == "trains"
},
func(_ bool) fyne.CanvasObject {
    return widget.NewLabel("Template")
},
func(uid TreeNodeID, _ bool, template fyne.CanvasObject) {
    label := template.(*widget.Label)
    label.SetText(strings.Title(uid))
})
```

The preceding code will display a tree in the app. Once the second element has been expanded, this tree will look as follows:

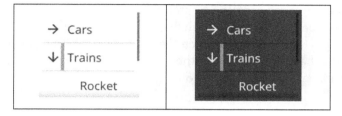

Figure 5.23 – Tree widget with the branch expanded, shown in the light and dark themes

The three collection widgets shown in this section provide useful functionality for presenting large or complex data. The API is a little more complex than the standard widgets, but this allows massive datasets to be presented to users; for example, by scrolling through thousands of records from a database or showing parts of a large file tree.

There's a selection of container widgets that we can use to build out more complex user interface designs and navigate through applications. We will discover these in the following section.

Adding structure with container widgets

In *Chapter 3, Window, Canvas, and Drawing*, we learned how a `Container` is used to group multiple objects within a canvas. Using the layouts we explored in *Chapter 4, Layout and File Handling*, it is possible to automatically arrange each `CanvasObject` according to certain rules. However, sometimes, an application would like items to appear and disappear according to user interaction, or to have visual attributes beyond their size and position. Container widgets can provide these richer behaviors. These structural widgets can be found in the `container` package and include scrolling, grouping, and variations of hiding and showing content. Let's explore each of these options (in alphabetical order).

AppTabs

The `AppTabs` container is used for controlling large areas of an application where the content should be switched out based on the current activity. For example, this may be used to fit lots of graphical elements into a small application user interface when only sub-sections are useful at one time.

Each tab in a tab container can contain text and/or an icon (whichever combination is used should be consistent for all items). Each tab has an associated `CanvasObject` (usually a container) that will be shown when the tab is selected. These are created using `TabItem` objects that have been passed to the `NewAppTabs` constructor function. To create two tabs with icons and labels, you would use the following code:

```
container.NewAppTabs(
    container.NewTabItemWithIcon("Tab1", theme.HomeIcon(),
        tab1Screen),
    container.NewTabItemWithIcon("Tab2", theme.MailSendIcon(),
        tab2Screen))
```

The preceding code will render as one of the following containers:

Figure 5.24 – Two tabs with text and icons in an AppTabs widget using the light and dark themes

The preceding image shows the tabs in their default orientation. However, the tab container can show tabs on any one of the four edges. The `SetTabLocation()` function takes one of the `TabLocation` types; that is, `TabLocationTop`, `TabLocationBottom`, `TabLocationLeading` (normally the left-hand side) or `TabLocationTrailing` (after content – normally on the right-hand side). The following image shows how the tab's location can change the icon's layout:

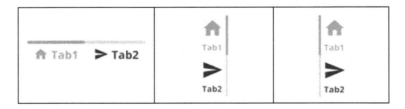

Figure 5.25 – Tab containers at the bottom, with leading and trailing locations

When running the application on a mobile device, it is expected for tabs to be on the top or bottom in portrait mode. Therefore, the locations will adapt appropriately – the leading setting will show the tabs at the top, while the trailing setting will stets the tabs at the bottom. If the mobile device is rotated, then the tabs will move to the left or right edge – leaving more space for content. In landscape mode, any tabs that have requested the top position will be shown on the leading (left-hand side) edge; the bottom setting will move to the trailing (right-hand side) edge.

Scroll

Most of the widgets that need to scroll content include the functionality to do so. However, if you want to add scrolling capabilities to your content, you can use the `Scroll` container. By wrapping some other element in a scroll container, you add scrollbars on the horizontal and vertical dimensions. The constructor function for scrolling in both the horizontal and vertical dimensions is `container.NewScroll()`. You can also call `NewHScroll()` if you would only like to scroll horizontally, or `NewVScroll()` if you would only like to scroll vertically. The following image shows full scrolling on simple label content:

Figure 5.26 – Scroll container showing a scrollbar and shadow in the light and dark themes

As you can see, the minimum size for a scroll container becomes very small – just 32x32. If you use the horizontal scroller, then its minimum height will fit the content, while if you use the vertical scroller (such as a list), then the width will adapt to fit the content.

Split

The `Split` container provides us with a neat way to separate two parts of our application when we would like our users to be able to change the amount of space available for each section. This can be split horizontally or vertically. A horizontal split container displays two elements side by side with a split bar between them. The vertical split will stack elements one above the other with a split bar in-between:

```
right := container.NewVSplit(
    widget.NewLabel("Top"), widget.NewLabel("Bottom"))
container.NewHSplit(widget.NewLabel("Line1/nLine2"), right)
```

In the following image, you can see a horizontal split container with `Line1\nLine2` on the left-hand (leading) side and a vertical split containing `Top` and `Bottom` on the right-hand (trailing) side:

> **Note**
>
> In horizontal mode, the leading position (first parameter) is normally on the left, while in vertical mode, it will be on the top.

Figure 5.27 – The Split widget in horizontal and vertical modes using the light and dark themes

The split container allows the bar to be dragged to change the size allocated to each side of the split. The container's minimum size will be the sum of the two contents (plus the split) and unless it is in a parent container with more space, the bar will not be draggable. When there is more space available, then dragging the bar will change where the extra space is allocated.

Developers can also manually specify the proportion directly using the `Offset` field. A value of `0.0` means that the split should be as far left (or up) as possible, while a value of `1.0` means it should be fully right (or bottom aligned). This value can be queried during app runtime if you want to save the user's preference.

As well as composing standard widgets together to form a clear and logical user interface, it is sometimes useful to display temporary information or request user input. For cases where developers or designers do not want to include this in their main interface, there is a package of standard popup dialog boxes we can use. We will explore this next.

Using common dialogs

During the user's journey of an application, you will often need to interrupt the flow to present information, ask the user for confirmation, or to pick a file or other input element. For this purpose, toolkits usually provide dialog windows, and Fyne does the same. Instead of opening a new window, the dialogs will appear over existing content in the current window (which works well across all platforms as not all manage multiple window applications well).

Each dialog has its own constructor function (of the `dialog.NewXxx()` form) that create the dialog to be shown later using `Show()`. They also provide a helper function to create and show it (of the `dialog.ShowXxx()` form). The last parameter of all these functions is the window that they should be displayed in. All the dialogs also support setting a callback when the dialog closes. This can be configured using the `SetOnClosed()` method.

In this section, we looked at the different dialog helpers that are available (in alphabetical order) before learning how to build a custom dialog for an application. Although, these will load in the current application theme, we only showed one image for each example.

ColorPicker

Fyne provides a standard dialog for picking a color within applications. This feature will present a selection of standard colors, a list of recently selected colors, and also an advanced area where specific colors can be chosen through value sliders, editing the channel values in **Red**, **Green**, and **Blue** (**RGB**) or **Hue**, **Saturation**, and **Lightness** (**HSL**), or by entering the RGB hex color notation directly.

The color picker can be created by calling `dialog.NewColorPicker()` and then using `Show()` or simply calling `dialog.ShowColorPicker()`. The parameters of the constructor are the title and message to be shown at the top, a callback function for when the color is selected, and the parent window to display within it:

```
dialog.ShowColorPicker("Pick a Color", "",
    func(value color.Color) {
        fmt.Println("Chose:", value)
    },
    win)
```

The preceding code will load up the picker, as follows:

Figure 5.28 – The simple color picker dialog

The previous image shows the default simple color picker. Advanced features are available if you want your developers to have more control.

Confirmation

The confirmation dialog allows you to ask a user to confirm an action. As well as providing the title and content for this confirmation, developers can pass a callback function that will be called when the user makes their decision, with the parameter being false for a negative answer or true for a confirmed one. Like all dialogs, the last parameter is the parent window:

```
dialog.ShowConfirm("Please Confirm", "Are you sure..?",
    func(value bool) {
        fmt.Println("Chose:", value)
    }, win)
```

The confirmation dialog will look as follows if the light theme is currently loaded:

Figure 5.29 – A confirmation dialog using the light theme

It's important to remember that showing a dialog does not stop the code that loaded it. The user's decision will be communicated through the callback; the rest of your code will continue uninterrupted.

File selection

As we saw in *Chapter 4*, *Layout and File Handling*, the `dialog` package can help with file selection – choosing which file to open or where to save content. Opening the **Open File** or **Save File** dialog follows the pattern of the other dialog widgets. In this case, the callback function will take a `fyne.URIReadCloser` or `fyne.URIWriteCloser` type and an error since these operations can fail for a number of reasons:

```
dialog.ShowFileOpen(func(read fyne.URIReadCloser, err error) {
    fmt.Println("User chose:", read.URI().String(), err)
}, win) {
```

The Open File dialog looks as follows:

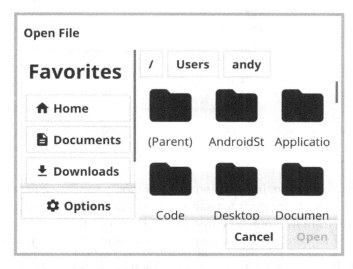

Figure 5.30 – The file dialog is used for choosing a file to open or save

The file dialogs will default to showing the user's home directory. This can be changed by calling the SetLocation method. As this is a cross-platform API, the starting location is a URI rather than a string path. This also means that the file dialog can be used to show the contents of remote file systems and other sources of file data.

In a similar way, applications can ask where to write a file to use the dialog. ShowFileSave method. It is also possible to prompt for folder selection instead of files, using dialog.ShowFolderOpen.

Form

The Form dialog extends the simple premise of a confirmation by requesting a value to be input, in addition to confirming the result. The Form dialog can contain various widgets in the same way that the Form widget did in the *Introducing the basic widgets* section. The constructor function is similar to the confirm dialog, but it accepts an additional slice of *widget.FormItem values to specify the content:

```
dialog.ShowForm( "Form Input", "Enter", "Cancel",
    [] *widget.FormItem{
        widget.NewFormItem("Enter a string...", widget.
            NewEntry())},
    func(bool) {}, win)
```

The result will be a dialog with an `widget.Entry` field, as shown in the following image:

Form Input

Enter a string...

✕ Cancel   ✓ Enter

Figure 5.31 – Asking a user for an input value using the Form dialog

Information

In a command-line application, information will commonly be written to standard output or standard error (normally for error messages). However, graphical applications will normally not be run from the command line, so messages that the user should see will need to be presented differently. The dialog package can help with this task as well.

An information dialog box can be used to present a standard message when its importance is high enough that the user should be interrupted to take a look at it. The `dialog.ShowInformation` function is called to present this dialog and it takes a title and message parameter. If the information to present is an error, then the `dialog.ShowError` helper function can be used as it takes an error type and the information is extracted to be displayed:

```
dialog.ShowInformation("Some Information",
    "This is a thing to know", win)

err := errors.New("a dummy error message")
dialog.ShowError(err, win)
```

The information dialog box will be presented like so:

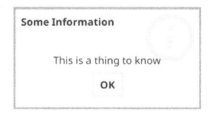

Some Information

This is a thing to know

OK

Figure 5.32 – An information dialog box in the light theme

After these, we move to the custom dialogs.

Custom dialogs

Although the preceding dialogs should cover most of the reasons why you may wish to interrupt the user flow with a pop-up dialog, your app may have additional requirements. To support this, you can insert any content into a custom dialog so that the overall layout is consistent.

To construct a custom dialog, a new parameter and its content must be passed to the constructor function. Any Fyne widget or `CanvasObject` can be used in a custom dialog, which includes containers to provide more complex content. To illustrate this, we will use a `TextGrid` component:

```
content := widget.NewTextGrid()
content.SetText("Custom content")
content.SetStyleRange(0, 7, 0, 14,
    widget.TextGridStyleWhitespace)
dialog.ShowCustom("Custom Dialog", "Cancel", content, win)
```

The preceding code will generate a custom dialog, as shown here:

Figure 5.33 – A dialog showing custom content (a TextGrid)

There is also a `ShowCustomConfirm()` version, which provides the **OK** and **Cancel** options. This behaves in the same way as the custom dialog shown in the previous image, except takes an additional `func(bool)` callback to inform the developer of which button was tapped.

By exploring various widgets and dialogs, we have seen what the standard theme looks like and that light and dark versions are available. Next, we will look at what a theme consists of and how they can be managed and customized.

Understanding themes

The themes within the Fyne toolkit implement the color palette, iconography, and size/padding values of the Material Design look and feel. The design of the theme API aims to ensure that applications feel consistent and deliver a good user experience while allowing developers to convey an identity and customization. All Fyne applications can be displayed in light or dark mode using built-in themes. We will look at these in detail next.

Built-in themes

Since more and more operating systems are supporting light versus dark desktop coloring, the Fyne theme specification supports both light and dark variants. By default, every app will ship with a built-in theme that provides both light and dark variants. This theme was illustrated extensively in the *Introducing the basic widgets* section earlier in this chapter, but to see how this all comes together, take a look at the following screenshot of a Fyne demo application that showcases widgets:

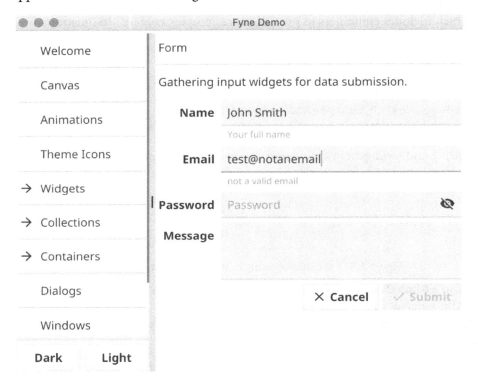

Figure 5.34 – A collection of widgets in a default theme – light variant

The previous screenshot shows the widget demo in the light theme. The following screenshot shows the same but with the built-in dark theme:

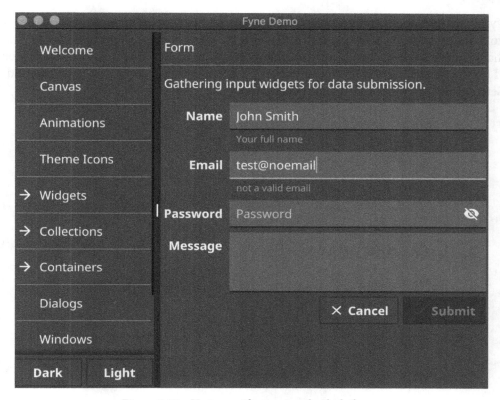

Figure 5.35 – Various widgets using the dark theme

As you can see, the main color being used (the primary color – blue, in this case) has been chosen as it contrasts well with the background colors of both the light and dark themes. When using this model, themes can vary the primary color while continuing to support both light and dark user preferences.

On most operating systems, Fyne will automatically pick the theme variant that best matches the current user's preferences. There are ways that the user can choose a specific version, as we'll see in the next section.

User settings

As we mentioned earlier, Fyne-based applications will normally detect the user's preference for either the light or dark theme and load it accordingly. It is possible to set a preference for which theme is loaded by using the Fyne settings application or by using environment variables.

The **fyne_settings** application, which can configure all Fyne-based applications, can be run to manage the user's settings. This includes their theme variant (light or dark), as well as what primary color they will use. Any changes that are made using this interface will be saved for the future and will immediately apply to all open applications. You can also find the **Settings** panel from the **Settings** menu within **fyne_demo**:

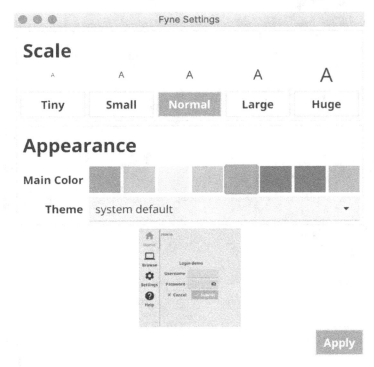

Figure 5.36 – The fyne_settings application

Using environment variables can be useful if you wish to temporarily apply a theme, or if you would like one app to use a different theme. The FYNE_THEME environment variable can be set to *light* or *dark* to specify which variant should be used. You can similarly override the default interface scaling that's available using the FYNE_SCALE environment variable. Here, 1.0 is the standard to use; smaller numbers load smaller content, while larger numbers load larger content.

Included icons

As could be seen in some of the widgets earlier in this chapter (for example, Button and AppTabs), the theme package includes many icons from the material design collection (the full official set can be found at https://material.io/resources/icons/).

Because all the elements of Fyne-based apps are designed to scale appropriately for different types of display and user preferences, the images should be vector-based rather than bitmap-based. This means that when displaying at very small or large sizes, the exact pixels to display will be calculated for optimal display instead of multiplying (or reducing) the number of pixels from the original image.

Thankfully, the material design images are available in vector formats and the built-in icons are all in **scalable vector graphics** (**SVG**) format. This also means that icons can easily be adapted for various colors as the app runs, thus ensuring that they can adapt to the primary color and current theme:

Figure 5.37 – A selection of material design icons

Because the icon set is freely available and very popular, it is easy to download additional icons and add them to your application, knowing that they will fit the toolkit's overall aesthetic.

Application override

Developers of applications that wish to deviate from the default (or user-selected) theme are also catered for in the Fyne theme API. Be careful before specifying a theme for your application – it may feel like a surprise to your app users. To use one of the built-in themes but override the user or system setting regarding whether the light or dark variant is used, you can call SetTheme() on the current App instance, as follows:

```
fyne.CurrentApp().Settings().SetTheme(theme.DarkTheme())
```

Alternatively, to force the application to use the built-in light theme, use the following code:

```
fyne.CurrentApp().Settings().SetTheme(theme.LightTheme())
```

This API is more commonly used to set up a custom application theme. The details of creating a custom theme will be covered in *Chapter 7, Building Custom Widgets and Themes*. Once you have created a theme, it can be loaded using the `SetTheme()` function, which will apply it to the current app. The following screenshot shows a custom theme that deviates from the standard styles:

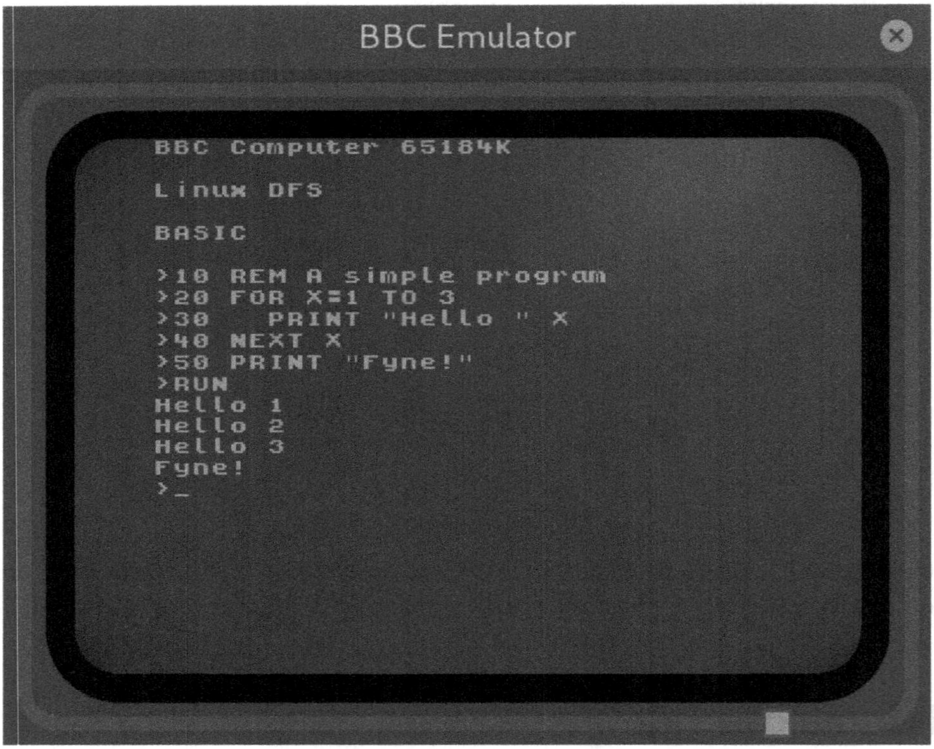

Figure 5.38 – A BBC Micro Emulator GUI based on Fyne

Now that we have explored the details of the main widgets and theme capabilities of the Fyne toolkit, let's build a simple app that brings many of them together.

Implementing a task list application

To explore some of the widgets listed in the previous sections and how they can be brought together into a simple application, we will build a small task list. This application will show a list of tasks based on complete or incomplete state and allow the user to edit the details of each item.

Designing the GUI

First, we will piece together a basic user interface for the task application. It will contain a list of tasks on the left-hand side of the app and a collection of components that edit a task on the right-hand side. Above this, we will add a toolbar for other actions. Let's get started:

1. The list of tasks will be a `List` widget that notifies the user when an item has been selected. The `List` widget will contain static content for this mock-up. Here, we will tell the list that there are a set number of items (for example, 5) so that it creates the correct number of items to display. We create a new check item each time the list calls the `CreateItem` callback. For now, we will leave the third (`UpdateItem`) method empty so that it just displays templates values. This code will be created in a simple `makeUI` method, as shown here:

```go
func makeUI() fyne.CanvasObject {
    todos := widget.NewList(func() int {
        return 5
    },
    func() fyne.CanvasObject {
        return widget.NewCheck("TODO Item x",
            func(bool) {})
    },
    func(int, fyne.CanvasObject) {})

    ...

}
```

2. Next, we will create the widgets that will allow us to edit a task item. Let's create a `Form` widget that will hold the items we need and provide labels as well. We will create a new row for each item using `widget.NewFormItem`, passing the `string` label and the widget's content as parameters. These are all standard widgets, but the callbacks we are passing are empty at the moment. We will return to these widgets later to complete their functionality. The following code goes inside the `makeUI` function we started in the previous segment:

```go
details := widget.NewForm(
    widget.NewFormItem("Title", widget.NewEntry()),
    widget.NewFormItem("Description",
        widget.NewMultiLineEntry()),
    widget.NewFormItem("Category",
        widget.NewSelect([]string{"Home"},
            func(string) {})),
    widget.NewFormItem("Priority",
        widget.NewRadioGroup([]string{"Low", "Mid",
            "High"},
```

```
                        func(string){})),
            widget.NewFormItem("Due", widget.NewEntry()),
            widget.NewFormItem("Completion",
                widget.NewSlider(0, 100)),
    )
```

3. The last component that we will add to the interface is a toolbar that will provide access to the add task function. To do so, we will create a widget.Toolbar using the ToolbarAction item:

```
    toolbar := widget.NewToolbar(
        widget.NewToolbarAction(theme.ContentAddIcon(),
            func() {}),
    )
```

4. To bring these interface elements together, we will create a new container using the Border layout. The toolbar will be set as the top item and the task items will be on the left of the container. Our form will take up the remaining space by being passed as a component that's not specified as being on a border. This container will be returned from the makeUI function so that it can be used to display our application window:

```
    return container.NewBorder(
        toolbar, nil, todos, nil, details)
```

5. To run our application, all we need to do is add the usual launcher code, which creates a window and sets our content. We do not need to specify a size for this window as the contents will naturally condense down to a sensible minimum size:

```
func main() {
    a := app.New()
    w := a.NewWindow("TODO List")

    w.SetContent(makeUI())
    w.ShowAndRun()
}
```

6. Running all the code we've created so far will give us a good impression of what the application will look like:

```
Chapter05$ go run .
```

When using the Fyne light theme (by going through the user preferences options or by setting FYNE_THEME="light"), the application should look as follows:

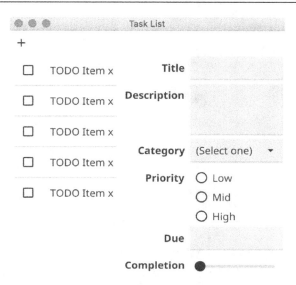

Figure 5.39 – Our task list GUI skeleton

Before we can complete the functionality of this application, we need to define a data structure that will hold information about the tasks we are editing.

Defining the data

For our application to function correctly, we will need to create a data structure that will manage the information we are editing. Let's get started:

1. First, we must define a task data structure – this simply lists the various fields that match the inputs in the design we made in the previous section. Different fields will be stored in different types – for example, the Entry widget maps to string and our checkbox maps to bool. We will add the following code to a new file called data.go:

```go
type task struct {
    title, description string
    done               bool
    category           string
    priority           int
    due                *time.Time
    completion         float64
}
```

As you can see, we have used float64 for the value of our completion Slider and that we will be converting the date entry into time.Time format.

2. Since we will be storing many tasks, we could simply create a slice of task pointers, but by defining a new type, we can associate certain functions with others that will be useful later. The type just wraps the `[] *task` slice, which will store the data:

```
type taskList struct {
    tasks []*task
}
```

3. Since we will be displaying the list of tasks based on the done state, we should add two helper methods that return these sub-lists based on the value of that field:

```
func (l *taskList) remaining() []*task {
    var items []*task
    for _, task := range l.tasks {
        if !task.done {
            items = append(items, task)
        }
    }

    return items
}

func (l *taskList) done() []*task {
    var items []*task
    for _, task := range l.tasks {
        if task.done {
            items = append(items, task)
        }
    }

    return items
}
```

4. We will also define some constants that help in managing the different priority levels in our data. Refer to the following code snippet:

```
const (
    lowPriority  = 0
    midPriority  = 1
    highPriority = 2
)
```

5. When writing data handling code, it's important to write tests as well. If you add these before connecting to the user interface, then bugs can surface sooner. This means that when we add graphical tests, any issues that are found should relate to a mistake in our user interface code. Create a new file called `data_test.go` and the following tests:

```go
func TestTaskList_Remaining(t *testing.T) {
    item := &task{title: "Remain"}
    list := &taskList{tasks: []*task{item}}

    remain := list.remaining()
    assert.Equal(t, 1, len(remain))
    done := list.done()
    assert.Equal(t, 0, len(done))
}

func TestTaskList_Done(t *testing.T) {
    item := &task{title: "Done", done: true}
    list := &taskList{tasks: []*task{item}}

    remain := list.remaining()
    assert.Equal(t, 0, len(remain))
    done := list.done()
    assert.Equal(t, 1, len(done))
}
```

More tests can be added to this project – you can find them in the code repository for this book, inside the `Chapter05` folder, at `https://github.com/PacktPublishing/Building-Cross-Platform-GUI-Applications-with-Fyne/tree/master/Chapter05`.

6. We have not explored data storage in this chapter, so we will just be keeping the task list in memory. You will find out more about data and preference storage in *Chapter 6, Data Binding and Storage*. Since our data will be reset each time the application is run, we should create another function that populates a data structure with some content that will be loaded when the app starts, as follows:

```go
func dummyData() *taskList {
    return &taskList{
        tasks: []*task{
            {title: "Nearly done",
                description: `You can tick my checkbox
and I will be marked as
done and disappear`},
```

```
            {title: "Functions",
                description: `Tap the plus icon above to
add a new task, or tap the minus
icon to remove this one`},
            }}
}
```

Now that we have defined the data structure and basic functions, we can connect it to the user interface and complete the functionality.

Selecting tasks

The simplest way to update a widget's content is to keep a reference regarding the instance once it has been constructed. We will be doing this for a number of elements, so we should create a new type that will handle the various elements of the user interface. Creating this struct means that we can avoid lots of global variables, which should help keep the code neat. Let's get started:

1. Create a new struct and name it taskApp, as follows:

```
type taskApp struct {
    data    *taskList
    visible []*task

    tasks *widget.List
    // more will be added here
}
```

    The type includes a reference to the *taskList data structure, which will hold our date, and defines a slice of *task types that represents the tasks that are currently visible (the result of calling taskList.remaining() or taskList.done()).

2. Now, we can make our makeUI function a method of the taskApp type so that its signature becomes func (a *taskApp) makeUI() fyne.CanvasObject. Doing this gives us access to the data structure we defined earlier through a.data. We will, however, use the task list stored in visible to populate our list as it may contain completed or incomplete items, depending on its current state.

3. The code that sets up the list widget can now be updated with the following code. We store its reference in a.tasks instead of the original todos variable (so that we can reference it later); don't forget to change the todos reference returned from makeUI to use a.tasks as well. The result of our Length callback function simply returns the number of items in the a.visible slice. Although the CreateItem callback (the middle parameter) does not need to change, we do provide an implementation for the final callback; that is, UpdateItem. This new function obtains the task from the specified index (i) and uses task.title to set the text of the Check widget:

```
a.tasks = widget.NewList(func() int {
    return len(a.visible)
},
func() fyne.CanvasObject {
    return widget.NewCheck("TODO item x", func(bool)
{})
},
func(i int, c fyne.CanvasObject) {
    check := c.(*widget.Check)
    check.Text = a.visible[i].title
    check.Refresh()
})
```

4. To see these changes in action, we need to set up the data source. For this, we must add a line that will create our dummy data and construct the new taskApp struct just before the call to SetContent, as follows:

```
data := dummyData()
tasks := taskApp{data: data, visible: data.
    remaining()}
w.SetContent(tasks.makeUI())
```

Performing these code alterations will update the app so that it reflects the task titles in the main list, as shown in the following screenshot:

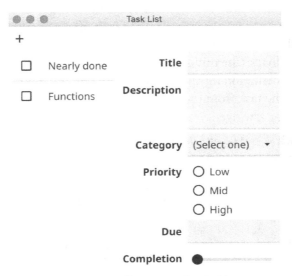

Figure 5.40 – Showing real task titles

Next, we need to fill in the details on the right-hand side of the window.

Filling in the details

To populate the data area of the app, we will need to keep track of the current task and the widgets that it should fill. Let's get started:

1. To do this, we will add a `current` field to the `taskApp` structure. After that, we need to save a reference to each of the input elements that we added for the initial layout tests, which will require more fields in `taskApp`:

```
type taskApp struct {
    data    *taskList
    visible []*task
    current *task

    tasks                   *widget.List
    title, description, due *widget.Entry
    category                *widget.Select
    priority                *widget.Radio
    completion              *widget.Slider
}
```

2. With these in place, we can complete the replacement for the `details` setup within `makeUI` so that it looks as follows:

```
a.title = widget.NewEntry()
a.description = widget.NewMultiLineEntry()
a.category = widget.NewSelect([]string{"Home"},
    func(string) {})
a.priority = widget.NewRadio(
    []string{"Low", "Mid", "High"}, func(string) {})
a.due = widget.NewEntry()
a.completion = widget.NewSlider(0, 100)

details := widget.NewForm(
    widget.NewFormItem("Title", a.title),
    widget.NewFormItem("Description", a.description),
    widget.NewFormItem("Category", a.category),
    widget.NewFormItem("Priority", a.priority),
    widget.NewFormItem("Due", a.due),
    widget.NewFormItem("Completion", a.completion),
)
```

3. Once this setup code is complete, we can add a new function called `setTask`. This will be used to update the current task and refresh the detail elements we created in the previous code block:

```
func (a *taskApp) setTask(t *task) {
    a.current = t

    a.title.SetText(t.title)
    a.description.SetText(t.description)
    a.category.SetSelected(t.category)
    if t.priority == midPriority {
        a.priority.SetSelected("Mid")
    } else if t.priority == highPriority {
        a.priority.SetSelected("High")
    } else {
        a.priority.SetSelected("Low")
    }
    a.due.SetText(formatDate(t.due))
    a.completion.Value = t.completion
    a.completion.Refresh()
}
```

4. To support that code, we will also define the `formatDate` function, which converts our date into a string value. This will return an empty string if the optional `date` is `nil`, or format it using the `dateFormat` constant otherwise:

```
const dateFormat = "02 Jan 06 15:04"

func formatDate(date *time.Time) string {
    if date == nil {
        return ""
    }

    return date.Format(dateFormat)
}
```

5. With this code in place, we can set the first task to be presented on the display. Of course, we should check if there are any tasks before assuming that an item can be shown. The following code is updated in the `main` function:

```
w.SetContent(ui.makeUI())
if len(data.remaining()) > 0 {
    ui.setTask(data.remaining()[0])
}
```

6. The last piece of code we need in order to update our user interface as the user browses is `List.OnSelected`. This will allow us to update the details that are displayed when the list is tapped. Simply add the following line once you've created our `List`, which is set to load from `a.tasks`:

```
a.tasks.OnSelected = func(id int) {
    a.setTask(a.visible[id])
}
```

With all the code in place, we have a complete application, as shown in the following screenshot:

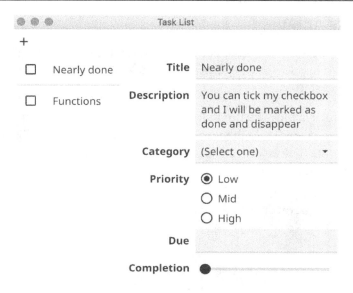

Figure 5.41 – Complete user interface

All the code shown previously will work with the current theme, which means we can see the same content when we're using the standard dark theme:

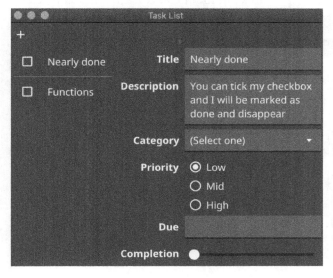

Figure 5.42 – The tasks user interface in the standard dark theme

Next, we will handle how details are saved when the user edits any data.

Editing content

When each of the input widgets are edited, we should update the dataset. This is trivial for most inputs as we can just set the OnChanged callback so that we're notified when the data changes. Let's get started:

1. In each callback, we must ensure that a task is currently selected (in case all the tasks have been deleted) and then set the appropriate field. The callback for Title is as follows. Note that we also call Refresh() on the task list as the title change should be reflected in the list:

```go
a.title.OnChanged = func(text string) {
    if a.current == nil {
        return
    }

    a.current.title = text
    a.tasks.Refresh() // refresh list of titles
}
```

Most of the other callbacks are similar, so they have been omitted from this description – the full code is available in this book's GitHub repository.

2. The priority callback update is a little more complex as we are converting a choice of string representations into a number field. Note that the callback is a function that's been passed to the constructor:

```go
a.priority = widget.NewRadio([]string{"Low", "Mid",
"High"}, func(pri string) {
    if a.current == nil {
        return
    }

    if pri == "Mid" {
        a.current.priority = midPriority
    } else if pri == "High" {
        a.current.priority = highPriority
    } else {
        a.current.priority = lowPriority
    }
})
```

3. Lastly, we will look at the input widget as we should add validation for the date format. To do this, we set the `Validator` callback to provide the user with feedback about the input state. First, we must create a new validator that can check the date format, which simply has a `Validate(string) error` function signature (meaning it implements `fyne.StringValidator`):

```
func dateValidator(text string) error {
    _, err := time.Parse(dateFormat, text)
    return err
}
```

4. With the validator in place, we simply set it as the `OnChanged` callback. In this callback, we need to reparse the date to get the appropriate date for the input (we skip this if the input is empty):

```
a.due.Validator = dateValidator
a.due.OnChanged = func(str string) {
    if a.current == nil {
        return
    }

    if str == "" {
        a.current.due = nil
    } else {
        date, err := time.Parse(dateFormat, str)
        if err != nil {
            a.current.due = &date
        }
    }
}
```

This is all the code we need for presenting and editing tasks. Next, we'll learn how to mark tasks as complete and keep the list updated.

Marking tasks as complete

The next piece of functionality we will add is the ability to mark a task as complete. Let's get started:

1. Because we are within the list, we need to set the callback within the `UpdateItem` callback to `List` to be able to mark the correct item as done:

    ```go
    check.OnChanged = func(done bool) {
            a.visible[i].done = done
            a.refreshData()
    }
    ```

2. Here, we need to make use of a helpful `refreshData()` function that updates the data list (by re-calculating what remains) and then asks the `task` widget to refresh:

    ```go
    func (a *taskApp) refreshData() {
        // hide done
        a.visible = a.data.remaining()
        a.tasks.Refresh()
    }
    ```

At this point, it functions correctly. However, upon clicking the `Check` text, it marks a task as done instead of selecting it for editing. To improve this, we can move the text component to a separate `Label` widget that will allow the mouse clicks through to the list selection logic.

3. To do this, we will return a `Check` and a `Label` by using `container.NewHBox` in the template function. When applying content in the update callback, we need to extract the widgets from the `Container.Objects[]` field. Otherwise, the code is similar to what it was earlier. The final list implementation is as follows:

    ```go
    a.tasks = widget.NewList(
        func() int {
            return len(a.visible)
        },
        func() fyne.CanvasObject {
            return container.NewHBox(widget.NewCheck("",
                func(bool) {}),
                widget.NewLabel("TODO Item x"))
        },
        func(i int, c fyne.CanvasObject) {
            task := a.visible[i]
            box := c.(*fyne.Container)
            check := box.Objects[0].(*widget.Check)
            check.Checked = task.done
    ```

```
        check.OnChanged = func(done bool) {
            task.done = done
            a.refreshData()
        }
        label := box.Objects[1].(*widget.Label)
        label.SetText(task.title)
    })
```

Finally, we will implement the add button in the toolbar.

Creating a new task

In this section, we will update the data code so that we can add new tasks. Let's get started:

1. First, we will create a new add() function with a task parameter and prepend it to the top of the list:

```
func (l *taskList) add(t *task) {
    l.tasks = append([]*task{t}, l.tasks...)
}
```

2. Because data functions are usually easy to test, we will add another unit test inside data_test.go:

```
func TestTaskList_Add(t *testing.T) {
    list := &taskList{}
    list.add(&task{title: "First"})
    assert.Equal(t, 1, len(list.tasks))

    list.add(&task{title: "Next"})
    assert.Equal(t, 2, len(list.tasks))
    assert.Equal(t, "Next", list.tasks[0].title)
}
```

Unit testing the whole user interface is highly advisable but outside the scope of this example – we will return to this topic in *Chapter 8, Project Structure and Best Practices*.

3. To complete the add task functionality, we must fill in the callback in the NewToolbarAction() function that we called when we first set up the user interface. This code simply creates a new task with the title New task, adds it to the data, and then reuses the same refreshData() function that we created for hiding completed tasks:

```
widget.NewToolbarAction(theme.ContentAddIcon(),
    func() {
```

```
        task := &task{title: "New task"}
        a.data.add(task)
        a.refreshData()
    }),
```

The preceding code concludes our tasks app example. There is more functionality that we could add, but we'll leave that as an exercise for you to complete.

Summary

In this chapter, we learned how the Fyne Widget API is designed and looked at a list of standard widgets. We saw how containers and collection widgets can help us organize and manage user interface components. The dialog package was also explored to show how we can use it with our applications in order to implement standard components for common activities.

We also saw how themes are implemented within the toolkit and how they apply to all the widget components. This chapter demonstrated the light and dark variants of the standard theme and showed that applications can provide their own themes for a custom look and feel.

By building a task tracking application, we saw how many of the built-in widgets are used, how to lay them out in various containers, and how user interactions can be tracked to manage some in-memory data. In the next chapter, we will look at data binding and storage APIs, which can help us manage more complex data requirements.

6
Data Binding and Storage

In the previous chapter, we learned that widgets can be controlled manually by application code. As we start looking at more complex applications, it's common for developers to want to display or manipulate a dynamic data source. The Fyne toolkit provides data and storage APIs that automate a lot of this work.

In this chapter, we're going to explore the ways that data is handled within the Fyne toolkit. We will cover the following topics:

- Binding data to widgets
- Adapting data types for display
- Binding complex data types
- Storing data using the Preferences API

By the end of the chapter, we will know how to make use of the data binding and storage APIs to create an app that helps track water consumption. It will store information on the local device and use APIs to explore how to minimize the coding required to manage data.

Technical requirements

This chapter has the same requirements as *Chapter 3, Windows, Canvas, and Drawing*, in that you must have the Fyne toolkit installed and a Go and C compiler working. For more information, please refer to the previous chapter.

The full source code for this chapter can be found at `https://github.com/ PacktPublishing/Building-Cross-Platform-GUI-Applications-with- Fyne/tree/master/Chapter06`.

Binding data to widgets

When we explored widgets in *Chapter 5, Widget Library and Themes*, we saw how information can be gathered and presented. Each widget we looked at was configured manually, and accessing data that the user entered (such as with the `Entry` widget) required code to query the widget's state. The `fyne.io/fyne/v2/data/binding` package provides functionality that supports automatically connecting widgets to data sources to handle this more efficiently.

In the following sections, we will explore what data binding is, why it is so useful, and how it is designed within the Fyne APIs.

Understanding data binding

There are many different approaches to data binding, and definitions can vary, depending on the toolkit that you are working with. In general, data binding allows the component of a graphical interface to have its display controlled by a separate data source. Moreover, it ensures that the graphical representation is always up to date with changes, and that user actioned changes within the user interface are synchronized back to the original data.

You can see how .NET approaches its implementation at `https://docs.microsoft. com/en-us/dotnet/desktop/wpf/data/data-binding-overview`. Android also provides similar functionality, which is documented at `https://developer. android.com/topic/libraries/data-binding`.

Despite the various approaches taken by each toolkit, the basics of any successful data binding are that the graphical output is automatically updated based on the state of a separate data object. This is known as **unidirectional** binding, often known as a *one-way data flow*. This is sufficient for data display when the information comes from an external source. However, it is not sufficient if your application will be modifying the data. Complete data binding (**two-way** or **bidirectional**) ensures that as well as keeping presented data up to date, it will also update the source data if the user alters the presentation through some interactive widget.

All the data binding that's done in the Fyne toolkit is bidirectional, meaning that each connection that's made can read or write the data it is connected to. Widgets that are only used for display purposes will not make use of the ability to write the data, but an input widget such as an entry or a slider will push data changes out to connected bindings. There's an underlying data source for each data binding. We will look at what data types are supported in the next section.

Supported data types

Just like the main Go language, all the values in the Fyne data API are strongly typed – this means that a data binding has a specific type of value. Each value matches a primitive Go type, ensuring that the compiler can check that these values are used in the correct way.

At the time of writing, the Fyne data binding supports the following types:

- `Bool`: A boolean value can be `true` or `false`. This type uses a `bool` for its internal storage.

- `Float`: A floating-point number value, this uses the `float64` type for internal storage.

- `Int`: A whole number value with positive and negative numbers, stored using the `int` type.

- `Rune`: A representation of a single unicode character, backed by the `rune` type.

- `String`: A bindable version of the Go `string` type.

- `List`: A mapping similar to `slice` or `array` that can contain a collection of a single type, indexed by an `int` offset.

- `Map`: A bindable version of the `map` primitive that can hold many types indexed by a `string` key.

- `Struct`: A data binding similar to `map` where the keys represent exported elements of a developer defined `struct` type.

The value type you use may be determined by the source data, or (if you are not bound to an existing data source) the output widgets you use. We will explore widget connections later in this section, but it is useful to know that we can convert types where needed. We will explore this later in the *Adapting data types for display* section.

When you have more complex data, such as a list or struct, you can still use the data binding API but with more advanced types, as discussed later in the *Binding complex data types* section.

Now that we've learned what data types are available, let's look at how to read and write data through a binding.

Creating, reading, and writing bound data

Each of the basic data types provided by the binding API (Bool, Float, Int, Rune, and String) provide two constructors: one that creates a new variable from the Go zero value (using the New... () name), and another that binds to an existing variable (named Bind... ()). It also provides Get () and Set () functions to manage data access. We'll explore these in detail next while using the Float type as an example.

NewFloat() Float

Creating a data binding using the New... constructor function will create a new piece of data with a standard value of zero. The returned object implements the DataItem interface, which enables binding. We'll look at this in more detail in the *Listening for changes* section, later in this chapter.

BindFloat(*float64) Float

The Bind... constructor function creates a bindable data item using a pointer to the primitive value. Using this function, we can set default values that are non-zero. Additionally, the original variable can still be used to get and set the data in places where data binding is not supported. In the following code, we are creating a new data binding to a floating-point value that defaults to 0.5:

```
f := 0.5
data := binding.BindFloat(&f)
```

The data variable that was returned is a new data binding of the Float type initialized to 0.5. If this binding is changed, it will write the new value back into the f variable.

Keeping Up to Date with Source Variables

If you are binding to variables that exist outside the data binding constructor, such as in the previous example, it will also cause changes from the binding to update the original variable. If, however, you update the variable directly, it is important to inform the data binding. Call data.Reload() to notify the data binding that the value has changed, so that it can propagate the change event.

Get() (float64, error)

To get the current state of a data binding, simply call `Get()`. It will return the contents in one of Go's primitive data types. For a `Float` binding, this will be `float64`. If an error occurred when you accessed this value, then an error will be returned in the second parameter. Though unlikely at this stage, you will see how bindings can be combined to create more complex scenarios. There is no additional action as a result of this call.

Set(float64) error

To update the value of a binding, you must call its `Set(val)` function while passing the new value (as with the `Get()` return value, this is a primitive data type). As a result of the value changing (if the new value is different to the current state), the data binding will notify all the code that is registered with it. If an error occurred when you were setting the value, the binding notifications will not be triggered, and the error will be returned.

Next, we'll look at how to request updates for changes that are occurring.

Listening for changes

One of the key concepts in data binding is that when a value changes, any item that is bound to it will update automatically. To do this, we need a mechanism that will notify us about changes. Before we learn how widgets do this automatically, we shall explore the implementation of this so that we can create our own code that stays up to date through data binding.

Using the DataItem interface to add a listener

The core functionality of any data binding is the `DataItem` interface. Each of the binding data types shown earlier (in the *Supported data types* section) also implements this interface. The definition of `DataItem` allows us to add and remove listeners. These listeners are informed whenever a data item changes:

```
type DataItem interface {
    AddListener(DataListener)
    RemoveListener(DataListener)
}
```

The definition shows how listeners are added and removed from any data item; this is how widgets can be informed of when data changes.

> **Current Values**
>
> When adding a listener to a DataItem, it will immediately be called with the current value stored. This enables any widget or program fragment that uses binding to be initialized correctly without additional code.

In this section, we saw that listeners are of the DataListener type. We shall explore what this type is in more detail in the next section.

Creating a DataListener

Each listener of a DataItem must implement the DataListener interface. By constructing a type that conforms to this definition, we can add, and later remove, the listener from the data items we depend on. It is defined as follows:

```
type DataListener interface {
    DataChanged()
}
```

As you can see, a single function is defined by the interface, DataChanged(). In most situations, the code that's using this will want to simply provide the function directly. For situations such as these, we can use a helpful constructor function that takes a simple function and returns an instance of a DataListener. The resulting listener can be used as a parameter of the AddListener and RemoveListener functions. It will call the function that was specified when the data is updated:

```
func NewDataListener(fn func()) DataListener
```

Using this knowledge, we can create a very simple data binding and watch for changes in its value. All we need to do is use one of the data binding constructors and call AddListener, passing the listener we created:

```
func main() {
    val := binding.NewString()
    callback := binding.NewDataListener(func() {
        str, _ := val.Get()
        fmt.Println("String changed to:", str)
    })
    val.AddListener(callback)
}
```

By running the preceding code, you will see that the callback is fired immediately with the current string value (in this case, the zero string, `" "`).

Depending on your computer speed, you may need to put a time delay at the end of the `main()` function as the application could exit before the data binding is processed (for example, `time.Sleep(time.Millisecond*100)`):

```
Chapter06$ go run listen.go
String changed to:
```

As shown in the preceding code, it is simple to create a data binding and be notified when the data changes. We could also trigger a change by calling `val.Set("new data")`, and the callback would trigger again.

Despite the preceding code being useful, the main usage for data bindings in Fyne is to connect widgets to data without having to write code that will copy information and watch for changes. Next, we'll learn how standard widgets operate with data bindings.

Using data with standard widgets

As we saw in the previous section, the data binding API allows us to create values that provide notifications when their content changes. As you might expect, the widgets provided in the Fyne toolkit understand bindings and can connect to them to display or update data. Let's explore how this can be set up:

1. We start by opening a new file in `package main` and creating a `makeUI()` function, as we have done previously. This time, we'll start the method by declaring a new floating-point number as our data binding. Using `NewFloat()` will create a new binding that defaults to the zero value of `0.0`:

    ```
    func makeUI() fyne.CanvasObject {
        f := binding.NewFloat()

        ...
    }
    ```

2. Next, we will create a `ProgressBar` widget that is bound to this data source. This progress bar simply displays a value that is a **one-way binding** – it only reads the data. To connect the bar to the `Float` data type we just created, we must use the `NewProgressBarWithData()` constructor function:

    ```
    prog := widget.NewProgressBarWithData(f)
    ```

3. To manipulate the data binding, we also want to include a **two-way binding** widget. For this example, we will use a `Slider`. As you saw in *Chapter 5, Widget Library and Themes*, this widget will display a value, using its current position, and allows the value to be changed by sliding. To set up this connection, we will use `NewSliderWithData()`, which takes additional parameters; that is, the permitted minimum and maximum values that can be sent into the data binding. We must then set the `Slider.Step` value (the gap between each step on the slider) to `0.01` so that we can cause fine-grained data changes:

```
slide := widget.NewSliderWithData(0, 1, f)
slide.Step = 0.01
```

4. In addition to the preceding one-way and two-way bindings, we can also use a version of one-way binding that is *write-only*. We do not need to connect the data to the widget in the same way we connected the last two widgets here (because there is no need for a listener for data changes). Instead, we can simply create a `Button` that will write to the data binding when tapped. We'll use the familiar `NewButton()` constructor function here and pass in the tapped handler, which will set the value to `0.5`:

```
btn := widget.NewButton("Set to 0.5", func() {
    _ = f.Set(0.5)
})
```

5. The final line in our `makeUI` function will be returning the container that packs these elements in. In this case, we will use `container.NewVBox`, which aligns each element on top of the others:

```
return container.NewVBox(prog, slide, btn)
```

6. To complete this example, we just need to create the usual `main()` function that creates a new `App` instance and open a window titled `Widget Binding`. The content from our `makeUI` function will be added and the relevant window will appear. Refer to the following code:

```
func main() {
    a := app.New()
    w := a.NewWindow("Widget Binding")

    w.SetContent(makeUI())
    w.ShowAndRun()
}
```

7. You can now run this example to see how these widgets behave when they're all bound to the same data source:

```
Chapter06$ go run widget.go
```

8. When the application appears, the slider and progress bar will be at the zero position. By dragging the slider, you will update the float that the slider and progress bar are bound to. As a result, you will notice that the progress bar moves. If you press the **Set to 0.5** button, it will set the value to **0.5**, and the other widgets will both update to the halfway position, as shown in the following screenshot:

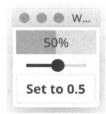

Figure 6.1 - Data bindings set to 0.5

In this section, we have seen the benefits of data binding for handling dynamic data, as well as the constraints of a strongly-typed API. Next, we will look at how to convert between types so that the data can be adapted for use in a wider variety of widgets.

Adapting data types for display

In the example we worked through in the previous section, we learned that it's possible to bind the same value to different widgets since both `Slider` and `ProgressBar` expect a `Float` value for their data. However, it is not always the case that these data types will align perfectly. Sometimes, we will need to perform conversions to connect to the widgets we wish to use. In this section, we will explore how to do so, starting with how we can include a label in the preceding example.

Formatting types into strings

In many applications, it is common to use a `Label` or other string-based display to contain information in another format, such as an `int` or `float64`. With data bindings, this is no different, so the `binding` package provides conversion functions that can make this adaptation easy.

To follow on from the previous example, we could include a `Label` that displays the `float64` value, but to do so, we would need a `String` binding rather than a `Float`. To obtain this, we can use the `FloatToString` function in the binding package. Like all binding conversion functions, this takes a single parameter, the source binding, and returns a new binding that is of the correct type:

```
strBind := binding.FloatToString(f)
label := widget.NewLabelWithData(strBind)
```

Using the string binding we just obtained, we can then create a `Label`, as shown in the previous code fragment. This approach will render the value using a default format. In the case of a `Float` binding, this will use the `"%f"` format string, which is similar to `fmt.Printf` in terms of usage. If we want to specify our own format, perhaps with some leading text, we can do so using the `ToStringWithFormat` notation instead:

```
strBind := binding.FloatToStringWithFormat(f,
    "Value is: %0.2f")
label := widget.NewLabelWithData(strBind)
```

When using this updated code, the output label will contain text such as `Value is: 0.50`, which is probably more meaningful to the users of your app. When specifying a format, be sure to include the appropriate format string for the source type, such as `%f` for `Float` or `%d` for `Int`.

Although displaying a value that is not a string is by far the most common conversion requirement, we may need to access a variable of some number type that is stored in a string. We'll explore this next.

Parsing values from the string type

In some cases, the data you are working with may not be in the desired format for the widgets you wish to use. When managing the conversion manually, you can do this as part of your own code – simply extract the data, convert it, and then supply the parsed information to your widget. However, with data binding, we want to keep a direct connection to the source data, which means conversion needs to happen within the chain of data binding.

Just like in the previous section, there are convenience methods that can help with this as well; they typically contain the phrase `StringTo` in their name. Let's work with an example that has a `string` value that contains an `int` number. We don't want to include manual code in the conversion, so we need to include a conversion in the chain, as illustrated here:

```
val := "5"
strBind := binding.BindString(&val)
intBind := binding.StringToInt(strBind)
```

As a result of this code, we have `binding.Int` that is reading the underling `string` data source and can now be used where a widget requires an `Int` as its data source.

This is a trivial example. Normally, when a number is stored as a string, it is because the value has some non-number element to it, such as a percentage that will include the trailing `%` symbol. When there is extra formatting in the string, we can still achieve the same outcome by using the `StringToIntWithFormat` version of the conversion, as follows:

```
val := "5%"
strBind := binding.BindString(&val)
intBind := binding.StringToIntWithFormat(strBind, "%d%%")
```

Notice that, in the format string, we needed to use a double percent to capture the percent symbol (`%%`). This is due to how the format strings work, but otherwise, this example is quite straightforward. Calling `intBind.Get()` will return 5 (with no error), and if the `strBind` value were to be changed to `25%`, then `IntBind.Get()` would return `25`, as expected.

In this section, we managed to use data binding conversion functions to change source data into a different type of data so that it can be used in our user interface. Next, we will learn how to make sure that changes in the output (for a two-way binding) are propagated to the original data.

Propagating changes through conversions

In the examples provided in the preceding section, we looked at complex chains of data binding so that we can convert source data for display purposes. This makes it much easier to use a variety of widgets connected to data binding without needing to write complex conversion code. However, we need to ensure that changes in the presented data are pushed back to the data source.

The great news here is that the conversions demonstrated in the *Formatting types into string* and *Parsing values from the string type* sections are full two-way data bindings. This means that we do not need to add any further code to propagate changes in the data. If an `IntToString` connection is made, then changes to the source `Int` will not only result in new values in the output `String`, but calling `Set()` on our `String` will cause a parsed integer to be pushed back to the original `Int` binding. This will depend on the string being correctly formatted, of course – setting an invalid value will not propagate a change it will return an error instead.

The types we've explored in this section are used by standard widgets to display and manage dynamic data. In many applications, however, we have more complex data. In the next section, we will explore how to bind such data to our apps.

Binding complex data types

The types that we have used in our data binding exploration so far have been limited to mappings of the Go primitive types. This means that they represent simple variables with a single element. For many applications, it will be necessary to display more complex data, such as lists, maps, or even custom structs. In this section, we will look at how that can be done, starting with the `DataList` type.

Using lists of data

Whether you wish to use data bindings to present data to a `widget.List` or a `widget.RadioGroup`, or if you are modeling your data with bindings that will be passed to your own widgets, the concept of a data list will be important. The data binding API defines `DataList` as a binding that provides additional `Length` and `GetItem(int)` functions, as follows:

```
type DataList interface {
    DataItem
    GetItem(int) DataItem
    Length() int
}
```

This generic definition means that a list can encapsulate a data type – the `DataItem` instance returned by `GetItem()` could be a `String`, `Int`, or even another `DataList`. You can implement a `DataList` using the previous interface or you can use one of the type-based lists provided by Fyne. The `Length` method will always return the number of items in this list, while `GetItem` can be used to access the data binding at the specified index (as lists are ordered).

Changes in the list's length (through `append`, `insert`, or `remove`) will trigger a callback on any listeners that are registered on the `DataList`. If the value of one of the items changes, it will trigger the change listener on the individual item rather than the containing list. Because of this, we can be clever about minimizing the impact of UI changes when a listener calls our code. We'll see how this process works in the following sections when we use a `StringList` to manage many items presented in a `List` widget.

Creating a list

Although a list can contain items of many different types, they will usually have all items of the same type, such as `String` or `Float`. For this purpose, there are helpful constructor functions, such as `NewFloatList` and `BindStringList`. You can build a new list with no content using the `New...List` methods, or you can bind to an existing slice in your Go code using `Bind...List` functions. For example, an empty string list might look as follows:

```
strings := binding.NewStringList()
fmt.Println("String list length:", strings.Length())
strings.Append("astring")
fmt.Println("String list length:", strings.Length())
val, _ := strings.GetValue(0)
fmt.Println("String at 0:", val)
```

Running this code will display the following output:

```
String list length: 0
String list length: 1
String at 0: astring
```

Here, you can see that the standard lists provide additional `GetValue` and `SetValue` methods, to access values just like the standard singular bindings. There are also matching `Get` and `Set` methods that allow you to change the whole list. This allows us to use the primitive types to access and apply data when the list type is known. We can also do the same using a slice for storage that the data binds to:

```
src := []string{"one"}
strings := binding.BindStringList(&src)
fmt.Println("String list length:", strings.Length())
strings.Append("astring")
fmt.Println("String list length:", strings.Length())
val, _ := strings.GetValue(0)
fmt.Println("String at 0:", val)
```

Running this code will result in the following output:

```
String list length: 1
String list length: 2
String at 0: one
```

Now that we have created a list binding, we can use it to display data in a widget that uses `DataList` bindings, such as `widget.List`.

Displaying a list

The most common way we will typically use a data list is through the `List` widget. When using a data binding to manage a list widget, it will automatically add rows that match the length of the data. It will grow or shrink when the length changes. Additionally, the data items in the list can be bound to the items in the list, meaning that if a data value is updated, the list will automatically update. Refer to the following code:

```
l := widget.NewListWithData(strings,
    func() fyne.CanvasObject {
        return widget.NewLabel("placeholder")
    },
    func(item binding.DataItem, obj fyne.CanvasObject) {
        text := obj.(*widget.Label)
        text.Bind(item.(binding.String))
    })
```

As you can see, we use the `NewListWithData` constructor function to set data binding in the `List` widget. The first parameter, `DataList`, replaces the `Length` function, which is used in a regular `List` constructor. The second parameter, the function that sets up template items, remains the same. The callback in the last parameter for updating items is similar to a regular list, except the first parameter is now a `DataItem` instead of a `ListItemID`. We can keep the `Text` value of `Label` up to date by calling `Bind()` and passing the `DataItem` once it has been cast to the `binding.String` strong type. Applying a binding in this manner will implement the standard API, so we don't have to worry about how caching will impact this functionality.

Managing lists with different types is possible, as long as you are careful when casting `DataItem` to the correct binding type on each access. Doing this is outside the scope of this chapter. In the next section, you will learn how the technique can be applied as a map of data, which typically contains many different types of values.

Using data maps

Maps of data in the context of data binding are much like the Go map type, where string keys map to DataItem types; for example, map[string]binding. DataItem. The DataMap definition is similar to DataList, except that it uses string keys to identify child elements instead of an int ID. Instead of the Length() method, which returns how long the list is, it requires a Keys() function, which returns a string slice containing all of the keys present in the dataset. To find the number of items present in the data, you can simply call len(DataMap.Keys()):

```
type DataMap interface {
    DataItem
    GetItem(string) (DataItem, error)
    Keys() []string
}
```

When an item is added or removed from DataMap, its listeners are fired. If the value of one of the items changes, it will trigger the change listener on the individual item rather than all of the DataMap listeners (as with items in a list, as mentioned in the previous section). Because maps normally have many different types of data in them, creating one using the standard API is a little different to creating a list, as we will see now. Of course, any type that implements the DataMap interface can be used in map data binding, but it is normally easier to use the provided implementations.

We will explore how DataMap can be used in the following sections by creating one and setting its values before describing how they could be used for display.

Creating a data map

DataMap, unlike DataList, does not have type-specific implementations within Fyne. Since maps normally contain many different types, there is only one map implementation that maps string keys to interface{} values. This can be created as a new map in memory, or by binding to an existing map with the map[string]interface{} type signature. First, let's look at creating a new map; the following code creates a new map using the binding.NewUntypedMap() constructor function and adds some values:

```
values := binding.NewUntypedMap()
fmt.Println("Map size:", len(values.Keys()))
"_ = values.SetValue("akey", 5)
fmt.Println("Map size:", len(values.Keys()))
val, _ := values.GetValue("akey")
fmt.Println("Value at akey:", val)
```

Executing the preceding code will produce the following output:

```
Map size: 0
Map size: 1
Value at akey: 5
```

We can do the same again but using an existing data source by making use of `binding.`
`BindUntypedMap()` instead, as follows:

```
src := map[string]interface{}{"akey": "before"}
values := binding.BindUntypedMap(&src)
fmt.Println("Map size:", len(values.Keys()))
_ = values.SetValue("newkey", 5)
fmt.Println("Map size:", len(values.Keys()))
val, _ := values.GetValue("akey")
fmt.Println("Value at akey:", val)
```

Executing the preceding code will produce the following output:

```
Map size: 1
Map size: 2
Value at akey: before
```

The preceding samples do not check for data types, but if you are sure of the type
for a key, you can perform type assertions like with a list. For example `values.`
`GetItem("key").(Float)` would return a float data binding to the float64 stored at
the specified key. Let us now look at how this can be used for display.

Displaying a map

At the time of writing, there are no built-in widgets that make use of the `DataMap`
binding type. Its main usage is to allow your application to maintain many data bindings
in a single binding type.

It is expected that both the `Form` and `Table` widgets will add support for data binding
in the near future. You can check out the latest API updates on the developer website at
`https://developer.fyne.io/api/widget.html`.

Before we leave this section, we will explore one last trick – mapping a custom `struct` to
a `DataMap`.

Mapping structs to a data binding

When populating data in a `DataMap`, one of the common ways to do so would be by using a `struct` type. In this situation, there are a set of fields where a name maps to a value (as we've seen many times in this book already). As you can see, this definition matches nicely to the maps that we saw in the previous section. To save a lot of manual code, the data binding API provides a simple way to automatically create a `DataMap` from an existing struct variable.

Just like any other `binding.Bind...` method, we pass a pointer to the variable and it returns the bound data type. In this case, the Fyne code will use the names of each `struct` field as the keys of the map, and the values will be set to the variables of the struct. To be able to bind a field in a `struct`, it must be an exported field (the name must start with an uppercase letter). The following code demonstrates this principle:

```
type person struct {
    Name string
    Age int
}
src := person{Name: "John Doe", Age: 21}
values := binding.BindStruct(&src)
fmt.Println("Map size:", len(values.Keys()))
name, _ := values.GetValue("Name")
fmt.Println("Value for Name:", name)
age, _ := values.GetValue("Age")
fmt.Println("Value for Age:", age)
```

Running the preceding code will produce the following output:

```
Map size: 2
Value for name: John Doe
Value for age: 21
```

As you can see, the keys and values all work as expected, but with the benefit that widgets can observe changes that have been made to the data and keep their display updated.

Now that we have explored how data bindings can easily handle how dynamic data is displayed, we shall look at how to store user-generated data using the Preferences API.

Storing data using the Preferences API

It is a common requirement for applications to store many pieces of information, such as user configuration options, current input field contents, and a history of opened files. Using files to store this information would require additional code to format the information for storage; using a database would require additional servers or dependencies for an application. To help with this, Fyne provides a **Preferences** API, similar to those used by iOS and Android developers.

Data that is stored as Fyne preferences can be accessed by any code in an application using a specific string identifier, known as a **key**. Each **value** that is stored has a specific type, so developers do not have to handle any conversion or type checking. Any time that this data changes, it will be saved for future use.

In this section, we'll learn how to manage data using the Preferences API and see how we can avoid having to manually manage user data.

Get and set values

Each supported type (see the following section) provides functions so that we can get and set data of that type. We will explore this using a string. To read or write data using a string, we can use the `String()` function; to write a value, we can use `SetString()`.

To gain access to the preferences, we can use `App.Preferences()`. If you do not have access to the `App` instance that loaded the application, you can use `fyne.CurrentApp().Preferences()` instead, which will return the same reference. There is an additional requirement, however – each application must declare a unique identifier for use in storage. To do this, we must change the `app.New()` constructor function to `app.NewWithID()`, passing a suitable unique identifier. Typically, the unique ID will be in reverse DNS format, and must match the identifier you will use during distribution (see *Chapter 9, Bundling Resources and Preparing for Release,* for more details). For example, you can use `com.example.myapp` for testing purposes.

The following code snippet sets up an application with a (relatively) unique identifier and accesses the standard preferences:

```go
func main() {
    a := app.NewWithID("com.example.preferences")

    key := "demokey"
    a.Preferences().SetString(key, "somevalue")
    val := a.Preferences().String(key)
    fmt.Println("Value is:", val)
}
```

We can insert the preceding code into the usual `main()` function (for a more complete listing, please go to the *Implementing a water consumption tracker* section). Notice that we used a single `string` value for the key – this helps us avoid making mistakes if typing out the key for each access. Running it will produce the following output:

```
Value is: somevalue
```

This quick example demonstrates string access, but there are other types we can use as well. We'll look at some of these in the next section.

Supported types

It's possible to store any type of data using just the string methods illustrated, but the API is designed to help us avoid the complexities of formatting and parsing data in that way. Due to this, there are different types supported by the Fyne preferences code. At the time of writing, the supported types are as follows:

- `bool`: Stores a simple boolean (`true` or `false`) value.
- `float`: Numbers that need a floating-point value can be stored using a `float`.
- `int`: For whole numbers, use the `int` functions.
- `string`: As we used previously, a simple `string` value.

Each of these types follow the same naming convention that we saw in the previous code. For example, you could set an integer using `SetInt()` or get a boolean value using `Bool()`.

By using Go semantics, the values that are returned will have zero values if no item was previously stored. It is possible to set different defaults using fallback values.

Fallback values

For situations where the default value of a property should not be the standard zero value (defined by Go), each `Get...` function has a `WithFallback` version; for example, `StringWithFallback()`.

If we change the code in the previous example so that it just uses the get and fallback methods, we can see how they work:

```
key := "anotherkey"
val := a.Preferences().String(key)
fmt.Println("Value is:", val)
val = a.Preferences().StringWithFallback(key, "missing")
fmt.Println("Value is:", val)
```

Running this version of the code will produce the following output:

```
Value is:
Value is: missing
```

With these methods, we can handle data with sensible defaults and save changes for future runs of the application. Sometimes, we will need to remove old data; we can do that too.

Removing old data

Storing data for users is important, but so is the ability to delete it when requested. To do so, the Preferences API provides one final method, RemoveValue, that will do just that.

By adding the following code to the end of our previous example, the values that were set will be cleaned out, meaning that on the next run, you will see the default values if the application is started a second time:

```
fmt.Println("Removing")
a.Preferences().RemoveValue(key)
val = a.Preferences().String(key)
fmt.Println("Value is:", val)
```

The preceding code also prints out the value once it has completed, ensuring that the item has been removed from the preferences. Running all the code from this section together will produce the following output:

```
Chapter06$ go run preferences.go
Value is:
Value is: missing
Value is: somevalue
Removing
Value is:
```

With that, we have seen how we can trivially store and access elements of data to be used in our apps. However, the Preferences API becomes even more powerful when we combine it with the data binding API we saw at the beginning of this chapter.

Binding to preferences

With the `binding` package, which we focused on earlier in this chapter, we can create data bindings that are connected to preference storage instead of regular variables. This provides us with the huge benefit that any time the setting of a value is triggered, it will be stored, and when the application is started again, the previous value will be loaded.

To access this functionality, we can use the functions whose names start with `BindPreference`, such as `BindPreferenceString()`. There is one function for each of the types supported by the Preference API, as listed earlier in this section. Each of these methods accepts a string parameter, which is the key string that we used in the previous code excerpts. Code that wishes to continue using the Preferences API can continue to do so as before, but using these data bindings ensures that new values are pushed directly to the widget that the binding is connected to. The bindings that are returned from a preferences bind use the same types as other data binding APIs, so you can get the `string` value of a preference item through its binding using `Get()`, as you might expect:

```
data := binding.BindPreferenceString("demokey",
    a.Preferences())
val, _ = data.Get()
fmt.Println("Bound value:", val)
```

The preceding code will access the same preferences value but through the data binding framework, making it easy to keep widgets up to date with user preferences. The output will be as follows:

```
Chapter06$ go run preferences.go
Bound value: somevalue
```

These bindings can also be chained, as with the earlier definitions, which means you can obtain a `String` binding to a preference value that is an integer by doing the following:

```
binding.IntToString(binding.BindPreferenceInt("mykey", p))
```

It is also possible to have multiple widgets, connected to multiple data bindings, all read and write the same preference value. If you use the same key to create many bindings to a preference, then they will all stay up to date when the value changes. We will see this in action in the example that follows.

With that, we have explored the data binding and Preference APIs and how they can, individually or together, vastly reduce the amount of code required to manage data in an application. Let's utilize this knowledge and implement an example application that can help us track our daily water consumption.

Implementing a water consumption tracker

The APIs that we have explored in this chapter can be helpful for most applications. To learn how we can add preference storage to a simple application, we will explore another example project. This time, we will create a tracker that can track water consumption over 1 week.

Constructing the user interface

Before we start working with data binding APIs, we will construct the basic user interface. The aim is to put today's total in large text at the top of a window, with the date below. We will follow this with the controls to add water to the current total. Under this, we will add another section that shows the values for the current week. Let's get started:

1. We will start, as usual, by defining a `makeUI` function, which builds the user interface. To start, we will define the large label that will be used to show the total by setting the font to 42 points, center aligning it, and using the `theme` primary color:

    ```
    func makeUI() fyne.CanvasObject {
        label := canvas.NewText("0ml", theme.PrimaryColor())
        label.TextSize = 42
        label.Alignment = fyne.TextAlignCenter
    ```

2. Now, we need to create another, regular label for the date to be displayed:

    ```
    date := widget.NewLabel("Mon 9 Nov 2020")
    date.Alignment = fyne.TextAlignCenter
    ```

3. The next few elements add controls, that support adding a value to the current water consumption. This could be a button that simply adds a specific number (for example, *250 ml*), but to be more flexible, we will allow the user to specify an amount. To do this, we will create an `Entry` field that is pre-filled with 250. Then, we will add a helper `ml` label after it and define a new button, labeled `Add`, that will action this later:

```
amount := widget.NewEntry()
amount.SetText("250")
input := container.NewBorder(nil, nil,
    nil, widget.NewLabel("ml"), amount)
add := widget.NewButton("Add", func() {})
```

4. Before creating the history layout we define a helpful `historyLabel` function. This will create a new label that simply contains `0ml` and aligns it to the right. The data here will be added later.

```
func historyLabel() fyne.CanvasObject {
    num := widget.NewLabel("0ml")
    num.Alignment = fyne.TextAlignTrailing
    return num
}
```

5. The last content element we'll add is the history information. In this case, we can construct this using a grid container with two columns. On the left, we will show the day of the week, while on the right, we will show a label for the value:

```
history := container.NewGridWithColumns(2,
    widget.NewLabel("Monday"), historyLabel(),
    widget.NewLabel("Tuesday"), historyLabel(),
    widget.NewLabel("Wednesday"), historyLabel(),
    widget.NewLabel("Thursday"), historyLabel(),
    widget.NewLabel("Friday"), historyLabel(),
    widget.NewLabel("Saturday"), historyLabel(),
    widget.NewLabel("Sunday"), historyLabel(),
)
```

6. To return the result of our interface building from this function, we can create a new vertical box. Inside this box, we must stack the total `label`, `date`, a horizontal grid that aligns the `input` and the `add` button, and lastly a `Card` widget containing the history content, along with a header.

```
    return container.NewVBox(label, date,
        container.NewGridWithColumns(2, input, add),
        widget.NewCard("History", "Totals this week",
            history))
}
```

7. For this example to run, we must create the typical `main()` function. This time, we want it to create a window titled `Water Tracker` and set the `makeUI()` return to the content before showing it:

```
func main() {
    a := app.New()
    w := a.NewWindow("Water Tracker")

    w.SetContent(makeUI())
    w.ShowAndRun()
}
```

8. We can now run this example in the usual way from the command line:

```
Chapter06/example$ go run main.go
```

Running the preceding command should result in the following interface appearing (when using the light theme):

Figure 6.2 – The empty user interface

This interface looks suitable, but it does not do anything yet. In the next section, we'll start creating some functionality by binding the total label to a value that can be incremented using the **Add** button.

Binding data to the UI

To start making the user interface functional, we will use data binding so that the header section of the app manages a single integer value representing how much water has been consumed in a day. Follow these steps:

1. First, we need to declare a new binding variable that is of the binding.Int type. We must add the following line to the beginning of our makeUI function:

```
total := binding.NewInt()
```

2. Next, we will add an implementation of the button tap handling. To action the number increment, we must replace the old `func()  {}` function with the following function:

```
func() {
    inc, err := strconv.Atoi(amount.Text)
    if err != nil {
        log.Println("Failed to parse integer:" +
            amount.Text)
        return
    }

    current, _ = total.Get()
    _ = total.Set(current + inc)
}
```

This will parse an integer from the `amount.Text` field and add it to the total. It does this by calling `Get()`, which finds the current value, and then `Set()`, with the increment applied.

3. Next, we want to display the integer value in a label, with **ml** (milliliters) added to the end. To do this, we can add a conversion that formats an integer for presentation. The following line creates a new `String` binding that is based on the `Int` function we already created:

```
totalStr := binding.IntToStringWithFormat(total,
    "%dml")
```

4. Now that we have used `canvas.Text` to define our large, colored text, we must bind these values. However, there isn't a helpful `WithData` constructor function for this, so we must apply the binding values manually. Let's create a new `DataListener` that will be called when the values change. Inside the callback, we set the text and request a refresh:

```
totalStr.AddListener(binding.NewDataListener(
    func() {
        label.Text, _ = totalStr.Get()
        label.Refresh()
    }))
```

5. With these changes made, we can run the app to see the changes in action:

```
Chapter06/example$ go run main.go
```

The user interface looks the same when it is loaded, but when we tap the **Add** button, the total value will update, increasing by the number by the value in the entry field:

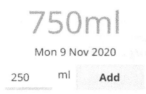

Mon 9 Nov 2020

250   ml   **Add**

Figure 6.3 – Showing a bound value

If you quit this version of the app and open it again, you will notice that the value is forgotten. We can fix this by using the Preferences API, which we will do now.

Storing with preferences

To remember values between application launches, we can make use of the Preferences API. This allows us to store values without needing to manage file access or databases. In this section, we will connect our data binding to the Preferences API so that it will automatically remember the data when it changes. Follow these steps:

1. To be able to store preferences, we need to allocate a key for each data item. In this example, each number is the total for a day, so we will use the date to identify the stored item. We need a new function, dateKey, that will format a string from any given time:

```
func dateKey(t time.Time) string {
    return t.Format("2006-01-02") // YYYY-MM-DD
}
```

2. To be able to use the Preference API, we need to make sure that our application has a unique identifier. To do this, we will use app.NewWithID instead of app.New. The ID should be globally unique. Then, we need to retrieve the application Preferences instance and pass it into our makeUI function so that it can be used in the bindings:

```
a := app.NewWithID("com.example.watertracker")
pref := a.Preferences()
w.SetContent(makeUI(pref)
```

3. Next, we will update the `Int` binding so that it can access preferences instead of creating a new value in memory. We will update `makeUI` so that it accepts a preferences instance and then change our binding creation so that it uses `BindPreferenceInt`. This constructor requires the key of the value it needs to access (which is generated by the `dateKey` function we created) and the preferences instance to use:

```go
func makeUI(p fyne.Preferences) fyne.CanvasObject {
total := binding.BindPreferenceInt(
    dateKey(time.Now()), p)
```

4. Since we are completing the top section of our app, we should also set the correct date. Simply change the label constructor so that it passes a formatted date string, like so:

```go
date := widget.NewLabel(time.Now().Format(
    "Mon Jan 2 2006"))
```

If you run the app again, you will see the same interface (with a corrected date display). If you add some value to today's total and then restart the app, you will notice that the value is remembered.

With that, we have completed the basic functionality of our app, but the history fields are still blank. Building on the previous functionality, we can show the stored values from previous days.

Adding history

With the main data binding code working, we can add similar functionality to the history panel. We will access a preferences value for each day of the week and update the `historyLabel` function to access it. Let's see how:

1. The most complex part of the history panel is date handling. We will need to find out what date represents the beginning of the week. Since the time of day does not normally matter (daylight savings are not accounted for in this code), we can just manipulate the date portion. Using `Time.Weekday()`, we can find what day of the week it is and go from there. Go's `time` package expects Sunday to be the first day of the week, so we need to handle this special case by subtracting 6 days to find Monday. For all others, we subtract the length of the days (`24 * time.Hour`) since Monday. The following code will allow us to find the date for our first history element:

```go
func dateForMonday() time.Time {
    day := time.Now().Weekday()
    if day == time.Sunday {
```

```
        return time.Now().Add(-1 * time.Hour * 24 * 6)
    }

    daysSinceMonday := time.Duration(day - 1)
    dayLength := time.Hour * 24
    return time.Now().Add(-1 * dayLength *
        daysSinceMonday) // Monday is day 1
}
```

2. Next, we must update the `historyLabel` function, adding the date of the day in question, along with a reference to the preferences to use. From this, we can generate our `dateKey`, as we did for the total storage, and bind an `Int` value to the preferences. Then, we create another string conversion using the same format, but in this instance, we can simply use the `NewLabelWithData` constructor function:

```
func historyLabel(date time.Time, p fyne.Preferences)
fyne.CanvasObject {
    data := binding.BindPreferenceInt(dateKey(date), p)

    str := binding.IntToStringWithFormat(data, "%dml")
    num := widget.NewLabelWithData(str)
    num.Alignment = fyne.TextAlignTrailing
    return num
}
```

3. Now, we can update the details of our history panel. First, we calculate the date for the Monday at the beginning of the current week, and then set a helper for the day's length, which is not part of the Go `time` package. For each history element, we pass the date for each day by incrementing by a number of days since the start of the week:

```
weekStart := dateForMonday()
dayLength := time.Hour * 24
history := container.NewGridWithColumns(2,
    widget.NewLabel("Monday"),
        historyLabel(weekStart, p),
    widget.NewLabel("Tuesday"),
        historyLabel(weekStart.Add(dayLength), p),
    widget.NewLabel("Wednesday"),
        historyLabel(weekStart.Add(dayLength*2), p),
    widget.NewLabel("Thursday"),
        historyLabel(weekStart.Add(dayLength*3), p),
    widget.NewLabel("Friday"),
        historyLabel(weekStart.Add(dayLength*4), p),
    widget.NewLabel("Saturday"),
```

```
        historyLabel(weekStart.Add(dayLength*5), p),
    widget.NewLabel("Sunday"),
        historyLabel(weekStart.Add(dayLength*6), p),
)
```

4. By running this final version of the app, we will see that it updates today's value in the history panel:

```
Chapter06/example$ go run main.go
```

The updated code will load the same interface that we had previously, but now, you will see that pressing **Add** will update the current day in the history panel as well:

Figure 6.4 – Our completed water tracker app

You will see the history on this app build up over a week, as indicated in the preceding screenshot. If you want to edit the data, you can modify the preferences file directly. The preferences file's location varies, depending on your operating system. More information is available at https://developer.fyne.io/tutorial/preferences-api.

Summary

In this chapter, we have looked at the various APIs available within Fyne for managing and storing data. We explored the concept of data binding and saw how it can help keep a user interface up to date, while at the same time reduce the amount of code we need to write.

We then looked at the Preferences API, which allows us to persist user data between application launches. When combined with the data binding code, this came with no additional complexity. By utilizing these features, we implemented an example application that manages data for tracking water consumption and stored it on our local device, ready to use the next day.

With that, we have covered the most common standard widgets and functionality in the Fyne toolkit. Sometimes, an application may require widgets or features that are not included. To support this, the toolkit allows us to extend the built-in components. We will explore this in the next chapter.

Building Custom Widgets and Themes

Over the course of the previous chapters, we have seen a lot of functionality that comes as part of the Fyne toolkit. Many applications will, however, benefit from components or functionality that are not included as standard. To be able to support an easy-to-use toolkit API and at the same time support additional functionality, the Fyne toolkit provides the ability to use custom code alongside regular widgets.

In this chapter, we will explore how custom data can be used in a Fyne app, and how custom styling can be added using code or by loading custom themes.

In this chapter, we will cover the following topics:

- Extending existing widgets
- Creating a component from scratch
- Adding a custom theme

At the end of the chapter, we will see how to make use of the custom widget and theme capabilities to create an app that presents a conversation history that could be used for various instant messenger protocols. It will demonstrate how a new widget can complement the standard set, and also how a custom theme can add some individuality to an application.

Technical requirements

This chapter has the same requirements as *Chapter 3, Windows, Canvas, and Drawing*: you will need to have the Fyne toolkit installed and a working Go and C compiler. For more information, please refer to the previous chapter.

The full source code for this chapter can be found at `https://github.com/ PacktPublishing/Building-Cross-Platform-GUI-Applications-with- Fyne/tree/master/Chapter07`.

Extending existing widgets

The standard widgets that we explored in *Chapter 5, Widget Library and Themes,* all have minimal APIs to define commonly required functionality. To support the addition of more advanced functionality, each Fyne widget can be extended by application developers. In this section, we will see how widgets can be enhanced by overriding their existing functionality or adding new behavior.

As we can see in the following diagram, extended widgets, as well as custom widgets, can be included in a container alongside standard widgets and canvas objects:

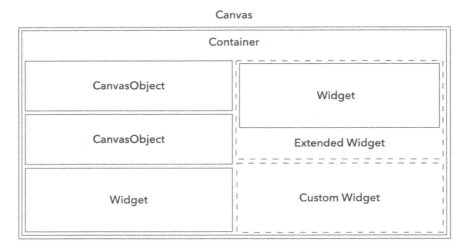

Figure 7.1 – Extended and custom widgets can be used alongside standard elements

An extended widget will embed an existing widget and provide replacement, or enhanced, functionality around it. Custom widgets, which we will see later in this chapter, implement the complete `Widget` interface, and so are not constrained by the designs of a standard widget.

Building custom widgets offers more flexibility; however, it requires a lot more code. Instead, we will start by learning how to extend existing widgets to add our own functionality. Once we have understood how to extend existing widgets, we will learn more about custom components in the *Creating a component from scratch* section.

Overriding widget functions

The first way that we will explore how to extend an existing widget is by overriding some of its functionality. This is normally accomplished by embedding a widget and creating a method of the same signature that Fyne would call into it, essentially replacing the built-in method with our own. When taking this approach, it is common to want the original functionality to execute after we have processed our override. To do this, our extended widget can call the original method from our own method.

To illustrate this, we will create an extended `Entry` widget that performs an action when the **Return** key is pressed. We call this `submitEntry`, and will enter the code in `submitentry.go`, as follows:

1. We first create a new struct that will define our custom type, naming it `submitEntry`. Inside it, we add an anonymous field, `widget.Entry`, which means that we will inherit all the fields and functionality of an `Entry` widget. Note that this is not a pointer:

    ```
    type submitEntry struct {
        widget.Entry
    }
    ```

2. Next, we create a constructor function, `newSubmitEntry`. This step is not strictly required, but it is essential that we call `ExtendBaseWidget()`, and so a function like this is usually the best approach. We need to pass the new widget as a parameter to `ExtendBaseWidget` so that the toolkit code knows that we are providing a replacement to the embedded widget:

    ```
    func newSubmitEntry() *submitEntry {
        e := &submitEntry{}
        e.ExtendBaseWidget(e)
        return e
    }
    ```

3. We can then add our own, overriding, functionality. In this case, we replace the
    TypedKey method, which is called when a key (physical or virtual) has been
    tapped to trigger an event. If we wanted to intercept characters, we would use
    TypedRune. In our method, we check whether the key is KeyReturn, and if it is,
    we take a custom action. If any other key is pressed, we call the TypedKey function
    of the embedded Entry widget (passing the same KeyEvent), which ensures that
    our widget continues to function as a text entry:

```go
func (s *submitEntry) TypedKey(k *fyne.KeyEvent) {
    if k.Name == fyne.KeyReturn {
        log.Println("Submit data", s.Text)
        s.SetText("")
        return
    }

    s.Entry.TypedKey(k)
}
```

4. Lastly, we create the usual main function. In this case, we simply set the content to
    our new submitEntry widget:

```go
func main() {
    a := app.New()
    w := a.NewWindow("Submit Entry")

    w.SetContent(newSubmitEntry())
    w.ShowAndRun()
}
```

5. We can now run the sample with the go run command:

```
Chapter07$ go run submitentry.go
```

You will see a window containing what looks like a regular entry widget, but if you hit the
Return key when typing, you will see a log message in the console and the content will
clear:

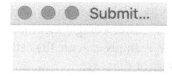

Figure 7.2 – The submitEntry struct looks like a regular widget.Entry

We have seen how to override an existing function of a widget, but it is also possible to add new features, as we will see when we learn how to make a tappable icon in the next section.

Adding new behavior

The second way that developers can extend existing widgets is to add new functionality by implementing new handlers around an embedded type. In this example, we will create a tappable icon. This extended `widget.Icon` will behave like a button with a single icon, but it does not include the border or tap animations of a regular button (which may not be desired in some situations). We start this example by creating `tapicon.go` and proceeding as described:

1. Once again, we start with a custom struct that embeds the existing widget, this time a `widget.Icon`. As before, this should not be a pointer type. In this struct, we will also include a `func()` to store the callback that should be run when the icon is tapped:

    ```
    type tapIcon struct {
        widget.Icon
        tap func()
    }
    ```

2. We create a constructor function again, primarily to ensure that `ExtendBaseWidget` is called. Into this we will pass a `fyne.Resource`, which specifies the icon to show, and a `func()`, which will be used to call when the icon is tapped. The resource is passed into the original icon as it will still handle the rendering:

    ```
    func newTapIcon(res fyne.Resource, fn func()) *tapIcon {
        i := &tapIcon{tap: fn}
        i.Resource = res
        i.ExtendBaseWidget(i)
        return i
    }
    ```

3. To add the tapped functionality, all we need to do is implement the `fyne.`
    `Tappable` interface, which requires a single method, `Tapped()`, taking a
    `*PointEvent` parameter. We simply execute the callback that was saved earlier
    inside this function, as long as one was set:

```go
func (t *tapIcon) Tapped(_ *fyne.PointEvent) {
    if t.tap == nil {
        return
    }

    t.tap()
}
```

4. For this demonstration, we will create a basic user interface that holds three of our
    `tapIcon` instances, simulating the home, back, and next navigation items from an
    app. To do this, we create a new `makeUI` function that aligns them in a horizontal
    box:

```go
func makeUI() fyne.CanvasObject {
    return container.NewHBox(
        newTapIcon(theme.HomeIcon(), func() {
            log.Println("Go home")
        }),
        newTapIcon(theme.NavigateBackIcon(), func() {
            log.Println("Go back")
        }),
        newTapIcon(theme.NavigateNextIcon(), func() {
            log.Println("Go forward")
        }),
    )
}
```

5. To complete this example, we create a new `main` function that will set the window
    content to the result of a `makeUI()` call:

```go
func main() {
    a := app.New()
    w := a.NewWindow("Navigate")

    w.SetContent(makeUI())
    w.ShowAndRun()
}
```

6. We can now run this whole example to see the resulting widgets in action:

```
Chapter07$ go run tapicon.go
```

When the application runs, you will see a window like the following that renders the icons. You can tap them and see the log output appearing when you do so. We have just recreated button widgets without the additional borders, padding, and tap animations:

Figure 7.3 – The tappable icons in a horizontal box

We have explored some different ways to extend existing widgets, but sometimes we want to create something completely new. In these situations, it is possible to build a widget from scratch, which we will do next.

Creating a component from scratch

Instead of building a new component by extending an existing widget, as we did in the previous section, we could build one from scratch. Any component that implements the `fyne.Widget` interface can be used as a widget in a Fyne application. To ease development, there is a `widget.BaseWidget` definition that we can inherit from. Let's start by defining the behavior of a new widget—the three-state checkbox.

Defining widget behavior

The API of a Fyne widget is based on behavior rather than how it looks. To begin our widget development, we will therefore define the states that our three-state checkbox can take and how a user can interact with it. We will create `threestate.go` and start coding:

1. Firstly, we must define a new type, `CheckState`, which will hold the three different states of our new checkbox widget. As we are building a reusable component, it is a good idea to export the types that are required, such as `CheckState` and the various states it defines. The usage of `iota` defines the first index and the following states will increment from that value:

```
type CheckState int

const (
    CheckOff CheckState = iota
```

```
        CheckOn
        CheckIndeterminate
)
```

2. We then define the core of the new component, calling it `ThreeStateCheck`, and setting it to inherit the basic widget behavior from `widget.BaseWidget`. Using `BaseWidget` is optional but it saves some coding. We add a field named `State` that will hold the current state of the check widget:

```
type ThreeStateCheck struct {
    widget.BaseWidget
    State CheckState
}
```

3. Next, we create a constructor function for this new type. As with previous examples, we need to call `ExtendBaseWidget`; in this case, the basic functionality that we have inherited is set up correctly:

```
func NewThreeStateCheck() *ThreeStateCheck {
    c := &ThreeStateCheck{}
    c.ExtendBaseWidget(c)
    return c
}
```

4. The last behavior element of this type is its ability to respond to tap events. We set up a `Tapped` handler, just as we did with the tappable icon in the previous section. This time, we will rotate the three states of this widget, wrapping to `CheckOff` if the previous state was `CheckIndeterminate`:

```
func (c *ThreeStateCheck) Tapped(_ *fyne.PointEvent) {
    if c.State == CheckIndeterminate {
        c.State = CheckOff
    } else {
        c.State++
    }

    c.Refresh()
}
```

That is all that we need to write to define the behavior of this new widget. However, because it is a new component (instead of an extension of an existing widget), we must also define how it will be rendered, which we will do next.

Implementing rendering details

For a new widget to be complete, it must also define how it will be rendered. This requires a new type that implements `fyne.WidgetRenderer`, as we will implement ahead. This new type must be returned from a `CreateRenderer` function on the widget implementation as well, as you will see in the code ahead. This renderer will use one of three checkbox icons—two are built into the Fyne theme and the third we will provide in this code base. Note that extra resources like this should be bundled for distribution, which will be discussed in detail in *Chapter 9, Bundling Resources and Preparing for Release,* in the *Bundling assets* section:

1. To start the renderer definition, we create a new type named `threeStateRender`; this should not be exported, as renderer details are private. This will hold a reference to the `ThreeStateCheck` that it is rendering, as well as a `canvas.Image` that will display one of the three icons used for our check widget:

```
type threeStateRender struct {
    check *ThreeStateCheck
    img   *canvas.Image
}
```

2. Just as if we were defining a layout for a container, we need to define how elements of a widget renderer are sized and positioned. In this example, we will simply specify that our checkbox icons should be set to `theme.IconInlineSize` to be consistent with other widgets. We define this as our `MinSize` and use the same value to size our widget when it is asked to define a `Layout`:

```
func (t *threeStateRender) MinSize() fyne.Size {
    return fyne.NewSize(theme.IconInlineSize(),
        theme.IconInlineSize())
}

func (t *threeStateRender) Layout(_ fyne.Size) {
    t.img.Resize(t.MinSize())
}
```

3. To complete our renderer, we must also define the additional methods: the
    `Destroy` (called when this renderer is no longer needed), `Objects` (which
    returns the list of graphical elements), and `Refresh` (which is called if a state
    changes) methods. These methods are relatively simple—most are empty, and the
    `Refresh` method simply calls a new `updateImage` method, which is defined in
    the next step:

```go
func (t *threeStateRender) Destroy() {
}

func (t *threeStateRender) Objects() []fyne.CanvasObject
{
    return []fyne.CanvasObject{t.img}
}

func (t *threeStateRender) Refresh() {
    t.updateImage()
}
```

4. The key to ensuring that the widget is up to date is the selection of the right image
    for the current state. We do this in a new `updateImage` method, described in the
    following code:

```go
func (t *threeStateRender) updateImage() {
    switch t.check.State {
    case CheckOn:
        t.img.Resource = theme.CheckButtonCheckedIcon()
    case CheckIndeterminate:
        res, _ := fyne.LoadResourceFromPath(
            "indeterminate_check_box-24px.svg")
        t.img.Resource = theme.NewThemedResource(res)
    default:
        t.img.Resource = theme.CheckButtonIcon()
    }
    t.img.Refresh()
}
```

It simply checks the state and picks a resource to display. In the normal states, we
can use built-in icons from the theme, but for our new indeterminate state, we
must load our own resource. As we noted earlier in this section, the asset should be
bundled, but we will explore this in more detail in *Chapter 9, Bundling Resources
and Preparing for Release*.

5. The last part of writing a `WidgetRenderer` is to return it from the
    `CreateRenderer` method defined on the widget that we have created:

```
func (c *ThreeStateCheck) CreateRenderer()
    fyne.WidgetRenderer {
        r := &threeStateRender{check: c, img:
            &canvas.Image{}}
        r.updateImage()
        return r
}
```

In this method, we set up the renderer and pass it a `canvas.Image` instance that it can
use to display. We then call the `updateImage` method that we defined earlier to ensure
that the initial state is rendered correctly:

1. To be able to run this demo, all we need to do is add the usual `main` function.
    This time, we will set the content to a single three-state checkbox using
    `NewThreeStateCheck()`:

```
func main() {
    a := app.New()
    w := a.NewWindow("Three State")

    w.SetContent(NewThreeStateCheck())
    w.ShowAndRun()
}
```

2. We can now run the code as usual with the `go run` command:

```
Chapter07$ go run threestate.go
```

Running the application will display the first window in Figure 7.4. Tapping the icon will
rotate the checkbox through the three states that are illustrated:

Figure 7.4 – The three states of our custom checkbox

We have now explored the various ways to extend widgets to add new behavior and to
create completely new components. As you can see from the figures in the chapter so far,
they all look similar because they respect the current theme. In the next section, we will
see how to make application-wide style changes using a custom theme.

Adding a custom theme

To bring some individual style or brand identity to an application, it can be useful to define a custom theme. This is a powerful feature, but should be used carefully. The selection of colors can significantly impact the readability of text elements and icons. Additionally, users of Fyne applications are allowed to choose between light and dark modes, so it is important that your theme reflects this choice where possible.

There are two ways that developers can create a custom theme—either by defining a new theme from scratch by implementing the Theme interface, or by inheriting from the standard theme. Each have their own benefits, and so we will explore both, starting with looking at creating a brand new theme.

Implementing the theme interface

All themes in Fyne are provided by implementing the fyne.Theme interface (much like any widget will implement fyne.Widget). The interface requires us to provide four methods that are used to look up the details of a theme. These methods are as follows:

```
type Theme interface {
    Color(ThemeColorName, ThemeVariant) color.Color
    Font(TextStyle) Resource
    Icon(ThemeIconName) Resource
    Size(ThemeSizeName) float32
}
```

As you can guess, these methods return the color, size, icon, and fonts provided by a theme. The four signatures vary, but the meanings are as follows:

- Color: The color lookup uses two parameters: the first is the name of the requested color (which we will explore more later) and the second is ThemeVariant. At the time of writing, there are two variants: theme.VariantLight and theme. VariantDark. This allows a theme to adapt to light and dark mode, although doing so is optional.

- Size: This lookup takes only the size name parameter. It is one of the ThemeSizeName constants, such as theme.SizeNamePadding or theme. SizeNameText.

- Font: This lookup takes TextStyle as its lookup parameter. A theme can choose which font to return for the various styles, such as TextStyle{Bold: true} for bold. The theme should also check the Italic and Monospaced fields. Other options may be added in future releases.

- `Icon`: The final method allows a theme to provide custom icons if desired. The parameter is the name of the icon resource and the returned resource can be an image in PNG, JPEG, or SVG format. It is normally advisable for you to use `theme.NewThemedResource` if returning an SVG file so that it will adapt to the theme variant.

The `Color` method is the most complex of these, as it is expected (but not required) to return different values depending on the `ThemeVariant` passed in. Most theme colors would likely change if the user switched from light to dark mode; you can see this in the standard theme. However, not all do. As the user is able to pick their preferred primary color, it is common to keep this consistent between modes.

To manage the various values for `ThemeColourName` and `ThemeSizeName`, the theme package provides collections of constants, named `theme.ColorNameXxx`, `theme.SizeNameXxx,` and `theme.IconNameXxx`. A complete theme should return a suitable value for each of these constants. At the time of writing, the size constants are `SizeNamePadding`, `SizeNameScrollBar`, `SizeNameScrollBarSmall`, `SizeNameText`, and `SizeNameInlineIcon`.

The list of colors is far longer, and will probably grow as new theming capabilities are added. The list currently required is as follows:

- `ColorNameBackground`
- `ColorNameButton`
- `ColorNameDisabledButton`
- `ColorNameDisabled`
- `ColorNameFocus`
- `ColorNameForeground`
- `ColorNameHover`
- `ColorNamePlaceHolder`
- `ColorNamePrimary`
- `ColorNameScrollBar`
- `ColorNameShadow`

Although it is recommended that you implement a return value (that adapts to the requested variant) for each of the colors and sizes listed, there may be new items added over time that your theme may not know about. To help adapt to these situations, it is possible to specify that the theme extends the built-in one (which will always have a suitable color available). We will look at this next.

Providing customizations to the standard theme

In some cases, an application developer may wish to only modify certain color elements of a theme, such as introducing their own primary color to match their company branding. To support this, you can implement a partial theme and ask that it delegate any items that are not defined to the default theme that ships as part of Fyne. To do this, you can partially implement a theme and call out to `theme.DefaultTheme()` methods to provide the standard values. This could also be used to change the font in an application, for example, while leaving the colors as standard.

Let's look at a simple theme customization that wants to use a monospaced font for all text and wants that text to be orange. We will start in a new file, `theme.go`, and begin as follows:

1. To implement the `Theme` interface, we need to define a new type. We will use an empty `struct` for now:

    ```
    type myTheme struct {
    }
    ```

2. To always use a monospace font, we can implement the `Font` function and return the default font resource for any request:

    ```
    func (t *myTheme) Font(fyne.TextStyle) fyne.Resource {
        return theme.DefaultTextMonospaceFont()
    }
    ```

3. We then want to specify that the text should be an orange color. To do this, we implement the `Color` method and return this custom value when the name is `Colors.Text`. We can ignore the `ThemeVariant` parameter as we are not providing different values for light and dark. By delegating to `theme.DefaultTheme()` for all other colors, we will specify that the default theme values should be used:

    ```
    func (t *myTheme) Color(n fyne.ThemeColorName, v fyne.
    ThemeVariant) color.Color {
        if n == theme.ColorNameForeground {
            return &color.NRGBA{0xff, 0xc1, 0x07, 0xff}
        }

        return theme.DefaultTheme().Color(n, v)
    }
    ```

4. We do not have any size considerations, but we must implement the method. We simply return the value from the default theme so that the current sizes will be used:

```
func (t *myTheme) Size(n fyne.ThemeSizeName) int {
    return theme.DefaultTheme().Size(n)
}
```

5. Similarly, we need to provide an empty Icon function that will return the default theme icon.

```
func (t *myTheme) Icon(n fyne.ThemeIconName)
    fyne.Resource {
    return theme.DefaultTheme().Icon(n)
}
```

6. To be able to demonstrate the theme, we create a simple interface with a Label, Entry, and Button. The following makeUI function returns these elements in a vertical box:

```
func makeUI() fyne.CanvasObject {
    return container.NewVBox(
        widget.NewLabel("Hello there"),
        widget.NewEntry(),
        widget.NewButton("Tap me", func() {}),
    )
}
```

7. Lastly, we create a main function that loads and runs our app. Note that this time we call App.Settings().SetTheme, which will set a new theme to be used, as shown in the following code:

```
func main() {
    a := app.New()
    a.Settings().SetTheme(&myTheme{})
    w := a.NewWindow("Theme")

    w.SetContent(makeUI())
    w.ShowAndRun()
}
```

8. We can now run this app in the usual way, or specify the dark theme as follows:

```
Chapter07$ FYNE_THEME=dark go run theme.go
```

And now we can see the result of our custom theme. All text is monospaced and of a bright orange color!

Figure 7.5 – Using a custom theme

Now that we have seen how app-specific themes and custom widgets can add to a Fyne application interface, we will bring it together in an example application. This time, we will build a chat app interface.

Implementing a chat app user interface

A common example of graphical applications, especially in a mobile context, is the messaging app. Although there are many messaging apps now, they often share the design of colored textboxes scrolling back through time. They are also either left or right aligned (with some padding for emphasis) to show incoming messages as distinct from outgoing. In this example, we will implement the message component to show text in this way and apply a custom theme to give the app an identity.

Creating a message widget

We start with the message widget that is used to display a single message. Each message will have a body of text and the name of the person who sent the message. Using the sender name, we can determine whether the message is outgoing. To begin, we define a custom `Widget` type that will hold this data in a new file, `message.go`:

1. To create a custom widget, we define a new type, named `message`, that extends `widget.BaseWidget`. We add to this our own fields, `text` and `from`, which will hold our widget state:

```
type message struct {
    widget.BaseWidget
```

```
      text, from string
}
```

2. We will also make use of some constant values in this example—myName is the
    name that we will use for outgoing messages. Obviously, in a real app this would
    be a user setting. messageIndent is a value that determines how much left or
    right space will appear in our message output to align the incoming and outgoing
    messages:

```
const (
    myName        = "Me"
    messageIndent = 20
)
```

3. As with our examples earlier in this chapter, we create a helpful constructor
    function that sets up the custom widget and ensures that ExtendBaseWidget is
    called:

```
func newMessage(text, name string) *message {
    m := &message{text: text, from: name}
    m.ExtendBaseWidget(m)
    return m
}
```

4. Most of the work in our custom message widget relates to its positioning and style,
    so its renderer is where we have to do most of the work. We start this by defining
    a custom renderer type, messageRender. There is no standard renderer type
    to extend, but we will want to save a reference to the message widget that it is
    rendering (in case we need to read its state). We also add Rectangle for the
    background and Label that will display our text:

```
type messageRender struct {
    msg *message

    bg  *canvas.Rectangle
    txt *widget.Label
}
```

5. An important part of `Widget` (or any `CanvasObject`) is to know its minimum size. This determines how layouts will pack the content on screen. Our size is complicated by the use of wrapped text—the available width will alter the height. We create a helper method, `messageMinSize`, that will return the actual minimum size for an available width, from which we subtract `messageIndent` to create the gap in the resulting display (making it clearer which messages are incoming compared to outgoing):

```
func (r *messageRender) messageMinSize(s fyne.Size)
    fyne.Size {
    fitSize := s.Subtract(fyne.NewSize(messageIndent,
        0))
    r.txt.Resize(fitSize.Max(r.txt.MinSize()))
    return r.txt.MinSize()
}
```

6. Now that we know the space required for the text, we can implement the `MinSize` method. We add `messageIndent` to the width so that the horizontal space is reserved:

```
func (r *messageRender) MinSize() fyne.Size {
    itemSize := r.messageMinSize(r.msg.Size())
    return itemSize.Add(fyne.NewSize(messageIndent, 0))
}
```

7. The main logic for our renderer is the `Layout` method. It must size and position the text and background rectangles within the `Widget`. All positions are relative to our widget's top-left position:

```
func (r *messageRender) Layout(s fyne.Size) {
    itemSize := r.messageMinSize(s)
    itemSize = itemSize.Max(fyne.NewSize(
        s.Width-messageIndent, s.Height))

    bgPos := fyne.NewPos(0, 0)
    if r.msg.from == myName {
        r.txt.Alignment = fyne.TextAlignTrailing
        r.bg.FillColor = theme.PrimaryColorNamed(
            theme.ColorBlue)
        bgPos = fyne.NewPos(s.Width-itemSize.Width, 0)
    } else {
        r.txt.Alignment = fyne.TextAlignLeading
        r.bg.FillColor = theme.PrimaryColorNamed(
            theme.ColorGreen)
    }
```

```
        r.txt.Move(bgPos)
        r.bg.Resize(itemSize)
        r.bg.Move(bgPos)
    }
```

After calculating the full size of the text content plus padding, we set up the graphical details of this component. If it is an outgoing message, we right-align the content and set it to a `blue` color; otherwise, we make it green. The calculated sizes and positions are then applied to the elements.

8. To complete the renderer, we must implement the remaining methods. These are mostly empty because this example does not use dynamic data. The `Objects` method returns each of the elements included in the order they should be drawn, so the background must be before the text:

```
func (r *messageRender) BackgroundColor() color.Color {
    return color.Transparent
}

func (r *messageRender) Destroy() {
}

func (r *messageRender) Objects() []fyne.CanvasObject {
    return []fyne.CanvasObject{r.bg, r.txt}
}

func (r *messageRender) Refresh() {
}
```

9. The last function to complete this widget is the method that links `Widget` with `WidgetRenderer`. We pass in the canvas objects that will be drawn to save us from recreating them later:

```
func (m *message) CreateRenderer() fyne.WidgetRenderer {
    text := widget.NewLabel(m.text)
    text.Wrapping = fyne.TextWrapWord
    return &messageRender{msg: m,
        bg: &canvas.Rectangle{}, txt: text}
}
```

This completes the custom component, but before we can test it, we need to create the user interface that will use them. We start by making a list of message widgets.

Listing messages

To create the rest of the user interface, we will create a new file, `main.go`, and add the standard components. Firstly, we create a list of messages:

1. Using the `newMessage` function we created earlier, it is simple to create a message list. We just create a `VBox` container and pass it a list of `message` widgets created using that helper function. Clearly, in a full application, this would use an external data source of some sort:

```
func loadMessages() *fyne.Container {
    return container.NewVBox(
        newMessage("Hi there, how are you doing?",
            "Jim"),
        newMessage("Yeah good thanks, you?", myName),
        newMessage("Not bad thanks. Weekend!", "Jim"),
        newMessage("Want to visit the cinema?", "Jim"),
        newMessage("Great idea, what's showing?",
            myName),
    )
}
```

2. We can implement a simple `main` function to show us our progress so far. This will be useful later when the full user interface is ready to run. For this version, we just set the window content to the list returned from `loadMessages()`. We give the window a sensible size and show it:

```
func main() {
    a := app.New()
    w := a.NewWindow("Messages")

    w.SetContent(loadMessages())
    w.Resize(fyne.NewSize(160, 280))
    w.ShowAndRun()
}
```

3. We can now run the message list to see the current work:

```
Chapter07/example$ go run .
```

The result is a list of messages, aligned according to sender, and displaying the appropriate color. This can be seen in the following diagram:

Figure 7.6 – Our messaging list in the default theme

This completes the message listing (we will add a scroll container around the list in the following section). We should also add an input section to send new messages, which we will also do next.

Completing the user interface

The message list looks great, but we want to be able to send new messages. Let's see how this can be done:

1. To add the remaining user interface elements, we create a new function, makeUI, and start by adding the loadMessages list:

```
func makeUI() fyne.CanvasObject {
    list := loadMessages()
    ...
}
```

2. We then create a footer that contains an Entry to capture the text and a Send Button to transmit the message:

```
msg := widget.NewEntry()
send := widget.NewButtonWithIcon("",
    theme.MailSendIcon(), func() {
    list.Add(newMessage(msg.Text, myName))
    msg.SetText("")
})
input := container.NewBorder(nil, nil, nil, send,
    msg)
```

From the preceding code, we can see that by using the `submitEntry` earlier in this chapter, you could also support the ability to send on return if you like. The button tap handler will create a new message widget and add it to the list, which will refresh. It then resets the text in our `Entry`. We return a `Container` called `input` that positions these elements appropriately.

3. Lastly, for this function, we return a new `Border` container that positions the input row at the bottom and uses the rest of the space for the message list. We also add a `Scroll` container around the list so it can contain more data than fits on the screen:

```
return container.NewBorder(nil, input, nil, nil,
    container.NewVScroll(list))
```

4. To use this new code, we update the `main` function to call `makeUI()` instead of `loadMessages()`:

```
w.SetContent(makeUI())
```

5. We can now run the app again to see the complete interface, as follows:

```
Chapter07/example$ go run .
```

6. This time, we can use the input box at the end to add a new message. In Figure 7.7, we added a **How about...?** message:

Figure 7.7 – Our messaging app in the default theme

This completes our application functionality, but we can still make it more interesting by applying a custom theme.

Adding some flair with a custom theme

The default look of Fyne applications is designed to be clear and attractive, but some apps may want to apply more identity or design flair. We will do this for our messaging app, starting with a new file, theme.go:

1. We start by defining a new type for our theme. It does not need to manage any state or inherit from other structs, so it is created empty:

```
type myTheme struct {
}
```

2. The main purpose of this custom theme is to return a different Color for certain elements. We start by adapting the Background color to display light or dark versions of a *blue-gray* color, depending on the current user setting (called ThemeVariant). We return the default theme lookup at the end of this function to show that we do not have a customization for other colors:

```
func (m *myTheme) Color(n fyne.ThemeColorName,
    v fyne.ThemeVariant) color.Color {
    switch n {
    case theme.ColorNameBackground:
        if v == theme.VariantLight {
            return &color.NRGBA{0xcf, 0xd8, 0xdc, 0xff}
        }
        return &color.NRGBA{0x45, 0x5A, 0x64, 0xff}
    }

    return theme.DefaultTheme().Color(n, v)
}
```

3. We will also provide a custom color for focused elements. We insert the following code into the switch statement in the previous code segment. Clearly, many more customizations could be provided if you desire:

```
    case theme.ColorNameFocus:
        return &color.NRGBA{0xff, 0xc1, 0x07, 0xff}
```

4. In this example, we don't want to provide custom size values or fonts, but you could return custom values if you want (as shown in the *Adding a custom theme* section earlier). We need to implement these methods, but we will return the default theme lookup so that the standard values will be used:

```
func (m *myTheme) Size(n fyne.ThemeSizeName) int {
    return theme.DefaultTheme().Size(n)
```

```
}

func (m *myTheme) Font(n fyne.TextStyle) fyne.Resource {
    return theme.DefaultTheme().Font(n)
}

func (m *myTheme) Icon(n fyne.ThemeIconName) fyne.
Resource {
    return theme.DefaultTheme().Icon(n)
}
```

5. To apply this theme to our app, we use `App.Settings.SetTheme()`. This should be called from the `main()` function before `ShowAndRun()`:

```
a.Settings().SetTheme(&myTheme{})
```

6. Once again, we can run this code and see the completed work:

```
Chapter07/example$ go run .
```

7. We will see the custom theme loaded. In Figure 7.8, it is running with `FYNE_THEME=dark` to show that our custom theme works for light and dark modes:

Figure 7.8 – Using the theme we just wrote in dark mode

We have seen how to implement custom widgets and themes to build an attractive messaging user interface. It is left as an exercise for the reader to actually send and receive messages through your favorite chat protocol!

Summary

In this chapter, we have seen how to deviate from the standard components and built-in theme in various ways. We explored how existing widgets can be extended and adapted, as well as how to build our own components from scratch. We also saw how custom themes can be created and how we can apply our own customizations to the default theme through theme extension.

With this knowledge, we created an application that was a mix of standard and custom components. We added some visual enhancements through our widget's renderer, but also created further customization by defining a custom theme. Through the code in this chapter, we learned how to customize individual elements and widgets, as well as how to make visual changes that apply across custom and standard widgets, using the theme API.

This brings us to the end of our exploration of the Fyne toolkit APIs and their functionality. In the following chapters, we will see how to create and manage GUI applications and how best practices can help make robust software that is easy to maintain. We will also learn that applications can be prepared for distribution and even uploaded to platform app stores and marketplaces. In the next chapter, we will explore the best practices around project structure and how to keep a growing GUI app robust and maintainable.

Section 3: Packaging and Distribution

Having explored the toolkit and built our first applications using Fyne, we see how easy it can be to build applications that work across multiple platforms. Now, it is time to look at how to manage our applications and prepare them for publication. As each distribution system has different requirements for metadata and packaging formats, this can be daunting, but the following section will walk you through the details. By the end of these chapters you will have packaged up your apps, installed them like native applications on your desktop and phone, and uploaded them to various app stores.

This section will cover the following topics:

- *Chapter 8, Project Structure and Best Practices*
- *Chapter 9, Bundling Resources and Preparing for Release*
- *Chapter 10, Distribution – App Stores and Beyond*

We start this section by reviewing how to organize our applications, and then move on to packaging and distribution.

8
Project Structure and Best Practices

The Go language comes with a well-understood set of best practices such as style, documentation, and code structure. Often, when applications start adding **graphical user interface (GUI)** elements, these best practices can be lost. Testing individual components and keeping a clean separation of types helps us maintain clean code that is easier to maintain over time. These concepts can be followed within GUI code as well, with support from a toolkit such as Fyne.

In this chapter, we'll explore how these concepts apply to graphical application development and how we can learn from them to make our GUIs easier to manage over time. We will cover the following topics:

- Organizing a well-structured project
- Understanding the separation of concerns
- Using test-driven development and writing tests for the whole application GUI
- Managing platform-specific code

Let's get started!

Technical requirements

This chapter has the same requirements as *Chapter 3, Windows, Canvas, and Drawing*; that is, you must have the Fyne toolkit installed and a Go and C compiler working. For more information, please refer to that chapter.

The full source code for this chapter can be found at `https://github.com/PacktPublishing/Building-Cross-Platform-GUI-Applications-with-Fyne/tree/master/Chapter08`.

Some parts of this chapter refer to managing platform-specific code, so it may be beneficial if you have two different operating systems available to work with.

Organizing your project

One of the design principles of the Go language is that you can start simple and build more structure into your project as it grows. Following this mantra, you can simply start a GUI project with a single `main.go` file inside a directory that's been created for the project. This will initially contain your entire application, starting from its `main()` function.

Starting simple

Once your user interface has grown from the very basics, it is a good idea to split it into a new file, perhaps named `ui.go`. Splitting the code in this way makes it clearer which code is simply booting an application (the `main()` function and helpers) compared to what is actually building the user interface.

By this time, you should be thinking about adding unit tests (if you have not already added them!). These tests will live in a file, alongside your code, that ends in `_test.go` – for example, `ui_test.go`. It is good practice to test all of your code, and for each new function or type you add, there will be new tests to ensure the code is working correctly over time. It is normal for there to not be a test file for the main function since its purpose is simply to wire up the application's components and launch it. A project that has passed the very basic stage might contain the following files:

```
project/
    main.go
    ui.go
    ui_test.go
```

This structure works well until an application needs to add some custom types. We will explore this next.

Adding new types

Since Fyne application user interface code focuses mainly on behavior, it is common to break up different areas of an application into separate areas, each defining its own type. Each type will define the data (or data access) that it represents, as well as the various methods that can operate on that information. These sections of the application code could be simple type definitions, where there will likely be a UI creation function, named makeUI() or something similar. Here's an example:

```
func (t *myType) makeUI() fyne.CanvasObject { … }
```

Alternatively, they may be custom widgets, in which case the type extends CanvasObject, so it can be passed into the wider GUI structure. In either situation, these types deserve a new file, such as mytype.go, and their own tests, in mytype_test.go. For example, an application that is growing may have the following structure:

```
project/
    editor.go
    editor_test.go
    files.go
    files_test.go
    main.go
    status.go
    status_test.go
    ui.go
    ui_test.go
```

This is a good structure for a simple application, but once the code base grows, especially if it contains libraries or supporting functionality that are not part of the GUI, you will likely want to consider using multiple packages. So, what happens then? Let's take a look.

Splitting code into packages

Packages are useful when you want to separate some complex code that manages, for example, data access or complex calculations from the user interface code that will display it. In these situations, a separate package allows you to maintain the clean divide and test these elements independently (see the *Understanding the separation of concerns* section later in this chapter). If you would like this subpackage API to be publicly available, then you can just create a new folder for it (for example, project/mylib). Alternatively, you could choose to group many subpackages under a standard pkg directory (that is, project/pkg/mylib).

However, if you would like to keep its API as an internal detail for this project, you can use the special `internal` package as its parent (that is, `project/internal/pkg/mylib`).

It can also be helpful to keep your code tidy if the GUI code makes use of the internal structure as well, so that the top of the project contains fewer files. It is common to use `project/internal/app/` for this purpose. So, an application that contains the `storage` and `cache` internal libraries might begin to look like this:

```
project/
    internal/
        app/
            editor.go
            editor_test.go
            files.go
            files_test.go
            status.go
            status_test.go
            ui.go
            ui_test.go
        pkg/
            storage/
                storage.go
                storage_test.go
            cache/
                cache.go
                cache_test.go
    main.go
```

Using this model, the library packages are all self-contained and the app package can depend on them to operate. The `main.go` file may depend on all of these packages to prepare the application and launch its GUI.

The preceding structure works well for a single application repository. The following command will install it directly (note that this should not include a scheme prefix such as `http://`):

```
$ go get projectURL
```

However, in some situations, you will require more than one executable in a project. We'll look at this in the next section.

Multiple executables

Regardless of whether the project is primarily a library or an application that contains multiple executables, there is a standard directory named cmd that can contain multiple subdirectories, one for each executable. Each package within cmd will contain the name of the application it should compile to, though the Go package will always be main so that it can be executed.

So, if your project was mainly a library but contained the mylib_gui and mylib_config executables, then you would have a structure like the following (omitting any internal code):

```
project/
    cmd/
        mylib_gui/
            main.go
        mylib_config/
            main.go
    lib.go
```

By using this format, someone could depend on your library by using the following command (without http:// being used):

```
import "projectURL"
```

They could also choose to install the GUI binary using the following command:

```
$ go get projectURL/cmd/mylib_gui
```

This flexibility allows repositories to have multiple purposes while maintaining a clean structure. Of course, there would probably be a lot of common code in internal/pkg/ or internal/app/ to allow this.

The preceding illustrations are just examples – each application will have different requirements and may wish to diverge from these common layouts. To help developers understand an application, its main() function should be at the root of a project or in a cmd/appname/ subdirectory. Similarly, a library's main API should be importable from the root of a project or the pkg/libname/ subdirectory. Following these hints will make your app or library easier to pick up for new developers who are familiar with the recommended Go project layouts.

In this section, we learned how an application can be split into multiple pieces. It may not have been clear why this is important, though. In the next section, we'll look at the separation of concerns, which shows how this approach helps keep our code clean and maintainable for the future.

Understanding the separation of concerns

As we mentioned earlier in this chapter, as well as when we discussed the Fyne toolkit API principles in *Chapter 2, The Future According to Fyne*, in the *Designing APIs for simplicity and maintainability* section, the concept of **separation of concerns** is important if we wish to maintain a clean code base. It enables us keep related code together without the fear of breaking other areas when we make a change.

This concept is closely related to the single responsibility principle, as introduced by Robert C. Martin in his *Principles of Object-Oriented Design* (http://www.butunclebob.com/ArticleS.UncleBob.PrinciplesOfOod) article. Here, he stated the following:

> "A class should have one, and only one, reason to change."
>
> –Robert C. Martin

In this respect, *concerns* have a wider scope than responsibilities in that they typically influence your application's design and architecture rather than individual classes or interfaces. Separation of concerns is essential in a graphical application if you wish to correctly detach your easily tested logic from the presentation code, which manages user interaction. By separating the concerns of an application, it is easier to test subcomponents and check the validity of our software without even needing to run the application. In doing so, we create more robust applications that can adapt to changes in requirements or technology over time.

For example, the Fyne widgets and APIs should not be incorporated into, or impact the design of, your business logic. Even a graphical API focused on behavior, such as the Fyne toolkit, should only be referenced by the presentation layer of your application (items in the project root or `internal/app` packages). It is for this reason that a robust application is split into multiple areas. Each of the supporting libraries will operate without any reference to the presentation layer or toolkit's capabilities. In this way, we keep the software open to change without it having a huge impact on unrelated areas.

In the next section, we'll learn how the provided test utilities help ease the creation of unit tests. We can do this by keeping our code separated into smaller components. This is useful for validating the behavior of the presentation code of our apps.

Test driving your development

The effort required to automatically test user interfaces or frontend software is often considered too expensive for the value it returns when it comes to avoiding future bugs. However, this is largely rooted in the toolkits being utilized or even the presentation technologies that have been chosen. Without full support for testing in the development tools or graphical APIs, it can be difficult to create simple unit tests without a huge amount of effort being needed.

One of the design principles of the Fyne toolkit is that the application GUI should be as easy to test as the rest of its code. This is partly made possible by the API's design, but this is further reinforced by the test utilities that we can provision. We will explore this later in this section. Using the following approaches, we will learn how a Fyne application can follow **test-driven development (TDD)**, even for the user interface components.

Designed to be tested

The modular design of the Fyne toolkit allows different drivers to be loaded for different systems or purposes. This approach primarily supports Fyne applications working on any operating system, without developers needing to modify their applications. An additional benefit of this approach is that an app can be loaded into a test runtime to execute various checks, without ever needing to display to the screen. This vastly improves the speed of test runs and also makes your tests more reliable (as user interaction cannot interfere with the test process).

By importing the `fyne.io/fyne/v2/test` package, we automatically create an in-memory application that is capable of creating virtual windows that contain actual application GUIs. These windows support the same APIs as regular windows so that your code can run the same as it ran previously. Each graphical element can be programmatically interacted with and tested to confirm its behavior and state, and even to verify its rendering output.

In the next two examples, we'll learn how to test the behavior of a user interface component and then how to verify that it rendered correctly, all without needing to display the GUI on a screen.

Testing our GUI logic

To test the functionality of an application, we must define a very simple GUI. It will have a `Hello World!` label, followed by an entry widget that we can use to specify our name. The final component – a simple button – will be triggered, updating the greeting based on the input. Here is how it works:

1. First, we define a simple `struct` called `greeter` that will hold references to these objects. For this example, we will write the following code to `ui.go`:

    ```
    type greeter struct {
        greeting       *widget.Label
        name           *widget.Entry
        updateGreeting *widget.Button
    }
    ```

2. Since we are not creating a custom widget in this example, we will define a small method named `makeUI` that will construct the widgets that represent this application. In this case, this is a simple vertical box container that combines all the widgets that we created following the preceding description. We create each of the widgets, assign them to variables of the `greeter` type, and then return a vertical box that packs them together:

    ```
    func (g *greeter) makeUI() fyne.CanvasObject {
        g.greeting = widget.NewLabel("Hello World!")
        g.name = widget.NewEntry()
        g.name.PlaceHolder = "Enter name"
        g.updateGreeting = widget.NewButton("Welcome",
            g.setGreeting)

        return container.NewVBox(g.greeting, g.name, g.up
            dateGreeting)
    }
    ```

3. To perform the update, when the button is tapped, we need an additional function, `setGreeting`, that will format a replacement string using `fmt.Sprintf`. It passes in the current content of the `name` entry widget to make the greeting personal. This looks as follows:

    ```
    func (g *greeter) setGreeting() {
        text := fmt.Sprintf("Hello %s!", g.name.Text)
        g.greeting.SetText(text)
    }
    ```

4. Lastly, we create a simple `main` function that will load the greeter, display it in a window, and run the application:

```
func main() {
    a := app.New()
    w := a.NewWindow("Hello!")

    g := &greeter{}
    w.SetContent(g.makeUI())
    w.ShowAndRun()
}
```

You can verify that the application works correctly by simply running it, as follows:

```
$ go run ui.go
```

In this section, we're focusing on the tests we must employ, so let's write a unit test that validates the behavior we defined earlier:

1. First, create the test file that will be provided along with this code; that is, `ui_test.go`. Here, we define a test with the standard signature. It must start with `Test` and accept a single parameter; that is, a `testing.T` pointer:

```
func TestGreeter_UpdateGreeting(t *testing.T) {
    ...
}
```

2. In this function, we will create a new instance of our greeter and request that the interface components are created. Once this has been run, we assert that the initial state is correct. Here, we use the `assert` package from the testify project by `stretchr` (more information can be found at `https://github.com/stretchr/testify`). This will use the `github.com/stretchr/testify/assert` import path, which should be added to the top of the file.

   By adding the following code, you can set up the user interface so that it can be tested and perform its first assertions:

```
    g := &greeter{}
    g.makeUI()
    assert.Equal(t, "Hello World!", g.greeting.Text)
    assert.Equal(t, "", g.name.Text)
```

3. The final step of writing this test is to execute the user steps and check the resulting changes. We're using Fyne's `test` package to simulate a user typing into the entry widget and then tapping the button to confirm this. After that, we confirm that the greeting text has been updated:

```
test.Type(g.name, "Joe")
test.Tap(g.updateGreeting)
assert.Equal(t, "Hello Joe!", g.greeting.Text)
```

With this test code, we can be sure that the user interface is working correctly. It can be simply executed like any other go test:

```
$ go test .
```

With that, we know that the app works correctly. However, it can be helpful to verify that the output is being rendered as expected. Let's write a new test to do that.

Verifying that the output is being rendered

In most situations, an application can be tested for correctness through behavior testing, as we saw previously. However, it is sometimes useful to actually see what will be rendered to check the result. If your application contains custom drawing code or complex layouts, this may be appropriate.

In this section, we will create a new test, similar to the one we created in the preceding section, but in this case, we will test the rendered output using another `test` utility. Let's see how this works:

1. Create a new method, as shown in the following code. Once you've set up the `greeter` type, pass `g.makeUI()` into `test.NewWindow()`. This will create an in-memory window that we can use to capture the output, as follows:

```
func TestGreeter_Render(t *testing.T) {
    g := &greeter{}
    w := test.NewWindow(g.makeUI())
}
```

2. With a test window created, we can get its content using `w.Canvas().Capture()`. This function will return an image with the interface rendered as if it were running in a real window.

   Now, we can use the `AssertImageMatches` test utility, which requires that the test compares this image to the named file, as follows:

   ```
   test.AssertImageMatches(t, "default.png",
       w.Canvas().Capture())
   ```

3. This code will compare the default look. Now, we can simulate user actions again and compare the new state to another image file with an appropriate name:

   ```
   test.Type(g.name, "Joe")
   test.Tap(g.updateGreeting)
   test.AssertImageMatches(t, "typed_joe.png",
       w.Canvas().Capture())
   ```

You can run these tests just like the behavior tests before, though this time, the tests will fail because the images it is being compared to don't exist. You will find two new files inside the `testdata/failed/` directory.

You should look at these files to see what is being drawn. If you agree that the output is correct, then these files can be moved to the `testdata/` directory. On a second run of these tests, you will see that they all pass as expected. The following screenshot shows what the `typed_joe.png` file looks like:

Figure 8.1 – The generated image from testing our user interface code

Such tests can be brittle since making a change to the design of the toolkit will cause them to fail. However, they can be helpful to highlight when a code change causes a graphical change that was unexpected. Therefore, when used appropriately in your test code, this approach can be a valuable addition to your validation process.

The tests we have explored in this section help verify that your application code is correct, and they should be run regularly. The best way to ensure this is to have a server run tests automatically. We will look at this next.

Continuous integration for GUIs

Continuous integration (**CI** – the regular way to merge a team's work-in-progress code so that it can be automatically tested) has become commonplace in software development teams. Adding this process to your team workflow is shown to highlight issues earlier in the development process, which leads to issues being fixed faster and, ultimately, better-quality software.

A critical part of this is automating the tests that exercise the entire source code – including the GUI. It is highly recommended to include not only a regular compilation of your entire application code, but also a full run of the unit tests upon each commit. Doing so will help you quickly identify breakages or unexpected changes in behavior. Various CI tools are available for this purpose, though looking at them is outside the scope of this book. These are helpful when you're configuring your automated processes as they ensure that tests such as the ones explored in this section are part of your regular testing and acceptance checks.

We have already seen that testing, and doing so regularly, is important, but how does this change if we want to have slightly different code for different platforms? At some point, most applications are likely to need system calls that vary based on the operating system. Next, we will look at how to do this while maintaining good code structure that is easy to understand.

Managing platform-specific code

Back in *Chapter 2*, *The Future According to Fyne*, we saw that the Go compiler has built-in support for the conditional inclusion of source files based on a system of environment variables and build tags. As an application adds more functionality, especially from a platform integration perspective, it is possible that the toolkit will not provide all of the functionality you are looking for. When this happens, the code will need to be updated to handle platform-specific functionality. To do so, we will use a variation of the conditional build – using well-named files instead of build tags. This is easier to read at the project level and should clearly indicate which files will be compiled for which platform.

Let's create a simple example: we want to read text out loud, but our code only has the ability to do so on macOS (Darwin). We will set up a simple `say()` function that does what we want in the `say_darwin.go` file:

```
package main

import (
    "log"
    "os/exec"
)
```

```go
func say(text string) {
    cmd := exec.Command("say", text)

    err := cmd.Run()
    if err != nil {
        log.Println("Error saying text", err)
    }
}
```

This simple function calls out to the built-in *say* tool, a command-line application bundled with macOS that allows text to be read out loud. Since this file ends with _darwin.go, it will only be compiled when we are building for macOS. To compile correctly when building on other platforms, we need to create another file that will be loaded instead. We will call this file say_other.go:

```go
// +build !darwin

package main

import (
    "log"
    "runtime"
)

func say(_ string) {
    log.Println("Say support is not available for",
        runtime.GOOS)
}
```

In this file, we must specify the build condition since there is no special filename format for all other platforms. Here, // +build !darwin means that the file will be included on any platform other than macOS. The method we'll be providing in this file simply logs that the feature is not supported. Finally, we must create a simple application launcher named main.go that will call the say() function:

```go
package main

func main() {
    say("Hello world!")
}
```

Running this code (using `go run .`) will read `Hello world!` out loud when it's run on a macOS computer. On other operating systems, it will print an error stating that the feature is not available:

```
Chapter08/say> go run .
2020/10/01 16:46:32 Say support is not available for linux
Chapter08/say>
```

We can handle platform-specific code in such a way that it's clear to anyone learning the project at hand and reading its code for the first time. Another developer could decide to add a `say_windows.go` file to add support for reading text on Windows. As long as they also update the build rules in `say_other.go`, the application will continue to work as expected but with the addition of Windows-based text reading. The benefit of this approach is that it does not require us to modify any of the existing code to simply add this new functionality.

Summary

In this chapter, we explored some of the tips and techniques for managing a GUI-based application written with Go. By carefully planning the modules of an application and how they interact, we saw that we can make any application easier to test and maintain. Since higher test coverage is a factor when it comes to increasing the quality of software applications, we looked at how we can use these techniques to test our graphical code, which is a notoriously difficult topic. We stepped through an example of writing test code for a simple GUI application that could be run automatically.

When it becomes necessary to adapt to a specific operating system, we need to learn how our code can adapt. With appropriate abstractions or by writing platform-specific code that is switched out by generic fallbacks, we can keep our applications easy to maintain, despite operating system differences.

Throughout this book, we have been running examples from the command line, next to their source code. This means that we have been able to incorporate files that exist in the current directory – but this is not going to be possible as we start to distribute our applications.

In the next chapter, we will look at how to include these extra assets (such as images and data files) in our application in preparation for their release.

9
Bundling Resources and Preparing for Release

Go applications are known for building simple application binary files that make them easy to install. However, the additional data required for graphical applications can make this challenging and has resulted in complex package formats and the introduction of installers as well. Fyne provides an alternative solution that allows apps to once again be distributed as a single file on most platforms.

Completing the packaging of an application requires metadata and an additional build step to prepare the files for distribution. This step allows applications to be installed to the local system or development devices alongside system-native apps, which we will study in this chapter.

We will walk through adding the various files an app will need at runtime. We will cover the following topics:

- How to include additional files in your application
- Checking for common **User Experience** (**UX**) mistakes to improve your GUI

- Packaging applications ready for distribution
- Installing on your computer or development mobile devices

At the end of this chapter, you will install your applications on your computer and smartphone for real-world testing.

Technical requirements

This chapter has the same requirements as *Chapter 3, Windows, Canvas, and Drawing*: to have the Fyne toolkit installed and Go and C compilers working. For more information, please refer to that chapter.

For deployment to Android devices, you will need to install the Android SDK and NDK (refer to *Appendix B, Installation of Mobile Build Tools*). To build for iOS devices, you will also need to install Xcode on your Macintosh computer (a Mac is required for licensing reasons).

The full source code for this chapter can be found at `https://github.com/ PacktPublishing/Building-Cross-Platform-GUI-Applications-with- Fyne/tree/master/Chapter09`.

Bundling assets

Go applications are designed to run from a single binary file. This means they can easily be distributed and do not rely on installation scripts. Unfortunately, this benefit results in a cost for developers—we cannot rely on resources being found along with our applications in the way that web or mobile app developers can (and as we have been doing during development). To ensure that our applications conform to this design, we must embed any required assets into the application binary. This includes fonts, images, and any other static content that is needed for the application to operate correctly.

The Fyne toolkit provides a tool for the bundling of assets that is recommended for any apps built using Fyne. The benefit of using this tool is that it generates `fyne.Resource` definitions for each embedded resource, which makes it easy to pass embedded assets into various Fyne APIs. This bundle tool is actually a command within the project's `fyne` command-line tool that is used in various examples within this book. The command is installed with a single `go get` command as follows:

```
Chapter09/bundle$ go get fyne.io/fyne/v2/cmd/fyne
```

The `bundle` command simply converts assets from the filesystem in to the Go source code, which can then be compiled into applications. This means that the compiled application will include the assets and therefore not rely on them being present on the filesystem when the app runs. The `bundle` command is a part of the `fyne` executable, and takes the file to embed as its main parameter. It prints the result to the system output, so we use console redirection (>) to send the generated Go source code to a suitable file, as shown in the following code snippet:

```
Chapter09/bundle$ ls
data
Chapter09/bundle$ fyne bundle data/demo.svg > bundled.go
Chapter09/bundle$ ls
bundled.go      data
Chapter09/bundle$
```

Once the file is generated, we can reference it using the created symbol (of type `*fyne.StaticResource`, which implements `fyne.Resource`). This can be used like any other resource, so we can load it as an image in the following way:

```
image := canvas.NewImageFromResource(resourceDemoSvg)
```

The generated variable name may not be ideal for your usage, but it can be changed using an additional command parameter. For example, if you wanted to export this new symbol, you could specify a simpler name that starts with an uppercase letter by adding the `-name Demo parameter`, as follows:

```
Chapter09/bundle$ fyne bundle -name Demo data/demo.svg >
bundled.go
```

The preceding command manages the inclusion of a single asset, but most apps will need to have many. Let's see how to add many resources.

Including multiple assets

In the previous example, we bundled a single file that included all the headers needed to make the bundle file a complete Go source file. To bundle multiple files in this way, we would need a new bundle file for each asset. This is probably not ideal and so the bundle tool includes a `-append` parameter that can be used to add more assets to the same bundle file.

To bundle a second file we use this new parameter and change the console redirection symbol to the append version (`>>`). For example, we can add `demo2.svg` to the same bundle output:

```
Chapter09/bundle$ fyne bundle data/demo.svg > bundled.go
Chapter09/bundle$ fyne bundle -append data/demo2.svg >>
bundled.go
```

The resulting `bundle.go` file will contain two definitions, `resourceDemoSvg` and `resourceDemo2Svg`.

In this manner, you can embed many resources, but it requires an additional command for each resource, which can be time-consuming and prone to human error. Instead, we can bundle all of the assets from a directory with a single command. To do so, we just use a directory path instead of the filename, using the same syntax as the first bundle we executed. The result of the following directory bundle will create the same output as running the two file commands shown previously:

```
Chapter09/bundle$ fyne bundle data > bundled.go
```

As you can see, it can be quite powerful to embed lots of data with a single command. The resulting file (`bundle.go`) should be added to your version control so other developers do not have to run this command.

To support this, an easy configuration works best when the assets are in a separate directory. Therefore, in addition to the file structure discussed in *Chapter 8*, *Project Structure and Best Practices*, it is common to add `data` directories alongside code that will utilize the embedded assets.

When the assets are updated, however, a developer may not remember the command to use, and so we will look briefly at how this can be automated.

Automating bundle commands

The Go compiler has a helpful `generate` subcommand that can be used to process resources such as the assets we have been bundling in this section. To make use of this tool, we add a `//go:generate` header to one of our source files (not the generated file, as this will be overwritten).

For this simple example, we add a new file named `main.go` that exists simply to include this header (normally there would already be a file available). Into this, we add a header line that tells Go how to generate our resources, before the package name:

```
//go:generate fyne bundle -o bundled.go data

package main
```

You can see here a change to how we were calling the command before—the inclusion of a `-o` parameter followed by the name we want to output to. This is introduced because within a `generate` command, we cannot use the command-line redirection tools that were in use previously. The parameter simply has the same effect – it specifies which file the output should be sent to. And so when we run `go generate` we see the same result as if we'd bundled the data directory manually. This is as shown here:

```
Chapter09/bundle$ ls
data      main.go
Chapter09/bundle$ go generate
Chapter09/bundle$ ls
bundled.go data      main.go
```

Using the preceding tools, we have prepared our application work without the asset files present, which makes it easier to distribute. Before we start packaging we should also check whether the Fyne toolkit has other tips for our app.

Checking for UI hints

As Fyne is built using **Material Design** principles, it is possible to make use of their recommendations for good and bad ways to use certain components and how you should and shouldn't combine elements for a great UX.

Built into the Fyne toolkit is the concept of **hints**. These are suggestions that widgets and other components can make about how an app could make changes to offer an improved user interface.

We will start exploring what these hints can offer by creating a simple example tab container application. This code snippet will load two tabs into a tab container (the `makeTabs()` function). We then include a `main()` function that will load a new app, create a window, and set the tabs to be its content. The function then runs our app in the usual way as follows:

```go
package main

import (
    "fyne.io/fyne/v2/app"
    "fyne.io/fyne/v2/container"
    "fyne.io/fyne/v2/theme"
    "fyne.io/fyne/v2/widget"
)

func makeTabs() *container.AppTabs {
    return container.NewAppTabs(
        container.NewTabItemWithIcon("Home", theme.HomeIcon(),
            widget.NewLabel("Tab 1")),
        container.NewTabItem("JustText",
            widget.NewLabel("Tab 2")),
    )
}

func main() {
    a := app.New()
    w := a.NewWindow("Tabs hints")

    w.SetContent(makeTabs())
    w.ShowAndRun()
}
```

With this code written, we can run it as normal. However, this time we will pass the additional `-tags hints` parameter to turn on the suggestions:

```
Chapter09/hints$ go run -tags hints .
```

When running, you will see the app as illustrated in *Figure 9.1*:

Figure 9.1 – The Tabs app looks like it's working correctly

There really are no surprises in how the app looks, but if you check the output of the application that's printed to the command line you will notice lots of output that we have not seen before. You will probably see something like the following (it may vary based on the version of Fyne being used):

```
Chapter09/hints$ go run -tags hints .
2020/10/07 14:06:08 Fyne hint:  Applications should be
 created with a unique ID using app.NewWithID()
2020/10/07 14:06:08   Created at:
 .../Chapter09/hints/main.go:18
2020/10/07 14:06:08 Fyne hint:  TabContainer items should all
 have the same type of content (text, icons or both)
2020/10/07 14:06:08   Created at:
 .../Chapter09/hints/main.go:11
```

As you can see in the preceding output, there are two lines marked `Fyne hint`. These are each the start of a new suggestion. The line after each instance is useful to show where in our code the hint refers to (the paths were partly removed for clarity). The preceding hints tell us the following:

- Our application is missing the unique ID that is required for some functionality to work – you will have learned about `appID` already if you read *Chapter 6, Data Binding and Storage*. We will discuss this further in the *Metadata, icons, and app IDs* section later in this chapter.

- The tab container that we created has a mix of tab styles; one has an icon and the other does not. We could resolve this by adding an icon to the `JustText` tab, or by removing the `Home` tab icon.

As you can see from this, it can be helpful to check the hints for your application. The small changes suggested by doing so can lead to an improved UX, or resolve future issues that have not yet been encountered.

As our app is now ready to be packaged, we need to think about how it will be presented in terms of its app name, icon, and other metadata.

Choosing metadata, icons, and app IDs

Before we start on the technical aspects of creating an application release, there are a few prerequisites to consider. The application name is probably set by now, but do you have a great description for it? Do you know how to articulate the key features of your software in a way that will grab the attention of potential users? Have you (or your design team) created a great app icon that will be memorable and somehow indicative of its functionality?

If you won't distribute your app through a managed channel, such as an app store or platform package manager, you should consider how the application will be discovered by your target audience. There is a lot of discussion and information online about **Search Engine Optimization** (**SEO**) and a growing amount about **App Store Optimization** (**ASO**) to be found, so we will not go into detail here. What is clear in the current software climate is that the ease of discovery and memorability of your app are now more important than ever before. The three most important aspects are the icon and description of the app, and the unique identifier that it will use in each store. We'll start by exploring the details of an app icon.

Application icons

Picking your icon is probably the single most important part of preparing an application for release. It needs to be memorable and also evoke some idea of what the software is for. A great icon should look good when displayed either large or small, and in general, tiny details should be avoided, or only used for unimportant aspects of the design. Make sure that your icon is created at a high resolution; using a vector format is advisable (for example, **SVG**), but if you are working with a bitmap format (such as **PNG**) then 1024 x 1024 pixels is the minimum requirement for an icon to look great on the widest variety of devices. It's also important to consider the use of transparency—depending on the platforms you wish to distribute to, this may or may not be recommended. Most desktop systems allow the use of shaped icons, but not all will allow semi-transparent areas and iOS does not allow transparency at all, whereas Android encourages it.

Take some time to look at popular or commonplace icons on each of the operating systems or desktop environments where you expect your application to be used. Can you match your icon style to each of them successfully? Does it seem like a particular shape or style will be expected by users of these systems? It may be best, or necessary, to create different versions of the graphic for different platforms. Doing so is not a problem, and can be accommodated by passing different icons to the build commands that we will study later in this chapter.

The packaging commands later in this chapter allow an icon to be specified; however, if you would like to set a default icon for your app, simply call it `Icon.png` or `Icon.svg`.

Describing your app

At this stage of development, you may have started to engage your audience and understand what they like about the software and who the target users will be. If not, then don't fear – just note that this is the time to think about how your description and supporting materials could best attract new users. Whether it's through a web search engine or an application marketplace, the text you use is critical for convincing anyone to install your application. As well as the name of the application and its main functionality, make sure you consider how it could benefit your users. What tasks do you expect they will be trying to complete when searching for the solution you have built? Don't worry about making this text long, but do try to include these important points.

How exactly you will ship your application is discussed further in *Chapter 10, Distribution – App Stores and Beyond*, but whether you intend to ship your application via an online store or a simple website, it's advisable to make sure you have completed your metadata before you continue to the release process. The information we have prepared here will be embedded in the packages we create and it's important that it retains consistency with the distribution metadata that will be used later in this chapter and beyond. User trust can be quickly lost and having an app icon that does not match the preview, for example, can cause concern. Remember that the description should match the name and the icon so that users will quickly recognize it once installed.

Application identifier (appID)

As we saw in the *Checking for UI hints* section, a unique identifier will be needed at some point for every Fyne app. If you have already used the preferences or storage APIs then this may already be set; however, if you have not then you need to pick the app ID at this stage as it is required to proceed to packaging your app on many operating systems.

An app identifier is used to uniquely recognize this software; as well as being globally unique, it must never change. Accidentally changing this would likely result in users losing their data and may also mean that updates are not sent to existing users of your software, so pick one now and be careful that it is kept consistent.

The normal scheme for choosing your unique ID is to use the reverse-DNS notation. This format will be familiar to developers who have worked with Java or Android packages, or Apple's **Uniform Type Identifier** (**UTI**). The format is based on the assumption that each developer, company, or product has a website or home page address that can be used as a namespace for their work. When such a grouping is applied then additional information can be used to identify the software component internally, making it a globally unique identifier. The *reverse* component of reverse-DNS is useful for sorting and searching, which is why it gained popularity in the management of software components.

The generic format is as follows:

```
<extension>.<domain name>.<optional categories>.<app name>
```

And so, following this format, an example company with domain name `myco.com` that is releasing a product named `tasks` in their category of `productivity` software might use the following app ID:

```
com.myco.productivity.tasks
```

The content of this string after the initial, reversed domain name can be whatever you choose; adding categories or another identifier is commonplace. It is not advisable, however, to add a version number as this string must remain identical for the life of your software to avoid some of the potential issues described previously.

If you do not have a website for your application, you could choose to use the location that it is stored in instead. It does not matter if you move the location in the future, as this is just an identifier – be sure to keep it the same if you do move the repository location. For example, a tutorial app stored on GitHub for user `dummyUser` might take the following app ID – notice that there is a third element to the domain name to remain globally unique:

```
com.github.dummyUser.tutorial
```

Now that we have our metadata in order, we can start packaging our application and then install it on our test devices.

Packaging applications (desktop and mobile)

To incorporate the metadata prepared in the preceding sections, we need to execute the *packaging* phase. This will take the standard Go application binary and attach or embed the required data based on the operating specifics. As each platform requires different data formats and produces different resulting file structures, we use the `fyne` tool once again to take care of the details.

Packaging for your current computer

To create a package from a Fyne project, we use the fyne package command. By default, this will create an application bundle or executable for the current operating system. When run on macOS this will create a .app bundle; on Windows it will be a .exe file (with additional metadata); on Linux it creates a .tar.gz file that can be used to install the app.

It is possible to build this for a different system as well, using the -os parameter, which we will explore later in this chapter.

Before packaging, it is a good idea to confirm that your application builds successfully using the go build command. When your app is ready, simply execute the fyne package command and it will process your app and metadata to create the platform-appropriate output. For example, on a macOS computer you would see the following:

```
Chapter09/package$ ls
Icon.png main.go
Chapter09/package$ go build .
Chapter09/package$ ls
Icon.png main.go  package
Chapter09/package$ fyne package
Chapter09/package$ ls
Icon.png    main.go     package      package.app
```

You can see that the go build command created a regular binary file, and that fyne package created an app bundle. When opened in macOS Finder, you can see how the icon has been applied to the output application bundle:

| Icon.png | main.go | package | package |

Figure 9.2 – The file icons from a macOS build

If you run the same commands on a Linux computer you would see the following:

```
Chapter09/package$ ls
Icon.png main.go
Chapter09/package$ go build .
Chapter09/package$ ls
Icon.png main.go  package
Chapter09/package$ fyne package
```

```
Chapter09/package$ ls
Icon.png    main.go      package      package.tar.gz
```

To read about installing the applications we have just built, you can skip to the *Installing your application* section. However, if you would like to prepare a build for mobile devices, read on, as we will do that next.

Packaging for a mobile device

As mobile apps cannot be created on the device, they have to be packaged from a desktop computer and then installed on the mobile device. We use the same tools as we used in the previous sections, with the additional -os parameter specifying ios or android as the target system.

As mobile apps require an app ID to build, we must also pass the appID parameter along with the unique identifier discussed in the *Application identifier (appID)* section earlier in this chapter.

Before packaging for iOS or Android devices you will need to install Xcode or the Android Developer Tools (discussed in more detail in *Appendix B, Installation of Mobile Build Tools*).

With Xcode installed on a macOS computer (due to Apple's licensing restrictions), you can build an iOS app using the following command:

```
$ fyne package -os ios -appID com.example.myapp .
```

To build an Android app package (.apk) use the following command:

```
$ fyne package -os android -appID com.example.myapp .
```

Now that you have your application bundle or binary file ready to be installed, we will see how to simply install your app on your desktop and mobile devices.

Installing your application

If you just want to install the desktop app on your computer or development devices then you can make use of the helpful install subcommand. There are two modes for the install tool, firstly to install on the current computer, and secondly to install on a mobile device that is set up for development.

Installing on your current computer

To install your application onto your current computer and make it available system-wide, you could simply execute the following:

```
$ fyne install -icon myapp.png
```

The icon file is the minimum required metadata for installing an app to the desktop. If you would like to avoid passing the -icon parameter each time, you can simply rename the file to Icon.png and it will be used by default. Once the application is installed, you will see it in your computer's program list with appropriate icons showing.

Installing on a mobile device

At this stage, we can install apps to a mobile device if it is set up for development.

> **Note**
>
> Development provisioning can be complicated and is out of the scope of this book.
>
> You can read more for iOS devices at https://help.apple.com/xcode/mac/current/#/dev5a825a1ca.
>
> For Android devices, you can read the documentation at https://developer.android.com/studio/debug/dev-options.

With a development-enabled mobile device, apps can be installed using the same install command by passing a -os parameter as either android or ios. For example, to install a generated .apk file on your Android device, use the following command:

```
$ fyne install -os android -appID com.example.myid -icon myapp.
png
```

As you can see, for the mobile app installation, we needed the additional appID metadata value. This value is passed into the package command that we explored in the preceding section. If the package is up to date, this value may not be required, but it's usually a good idea to pass it just in case.

And so, you can see that it's simple to install applications on the current computer or connected mobile devices. To make this possible, the Fyne tool was actually cross-compiling (that means compiling for a different type of computer). Let's now look into how that works in more detail.

Cross-compiling with ease

The ability to compile for different operating systems or architectures than the current computer is called **cross-compiling**. We saw it used in the previous section to package and install a mobile app from a desktop computer. By cross-compiling, we can also build applications from one computer for other types of desktop as well, for example using Windows to build a macOS application.

There are two ways that this can be done. Firstly, we will see how developers familiar with platform-specific compilation can use their normal tools to build for multiple platforms. After that, we will look at the `fyne-cross` tool and how it hides all of the complexity using a Docker image to manage compiling.

Using installed toolchains

When taking the traditional approach to cross-compiling, the computer will require an additional compiler **toolchain** for each platform and architecture that the developer wants to support. This is what provides the ability to compile the graphics and system integration code and typically comprises a C compiler and linker. The manner of installation for each toolchain varies depending on the current operating system as well as the target toolchain. Details for the various installations are available in *Appendix C, Cross-Compiling*.

With a toolchain installed, the build process is like regular Go cross-compiling where you specify GOOS and GOARCH environment variables to specify the target operating system and architecture. However, we additionally must specify CGO_ENABLED=1 (so that C integration is enabled) and also a CC environment variable that specifies which toolchain compiler to use.

A quick summary of the most commonly used compilers and the CC environment variable to use is as follows (for more information please see *Appendix C, Cross-Compiling*):

Target Platform	CC=	Download	Notes
macOS (Darwin)	o64-clang	github.com/ tpoechtrager/ osxcross/	You will also need the macOS SDK (see the linked osxcross page).
Windows	x86_64-w64-mingw64-gcc	macOS: brew.sh Linux: Use your package manager	The macOS package is mingw-w64. On Linux the name varies – usually mingw-w64-gcc, gcc-mingw-w64 or similar.
Linux	x86_64-linux-musl-gcc or gcc-linux	macOS: brew.sh Windows: tdm-gcc. tdragon.net	macOS package is FiloSottile/ musl-cross/musl-cross; on Windows, use the gcc package.

Table 9.1 – Downloads, notes, and CC environments for various desktop platforms

With the appropriate compilers and libraries installed, we can continue to the build phase. For each of the target operating systems, you will need to run through these steps with the correct environment variables set. It is recommended to build for one platform and then complete the packaging step for each before changing to the next configuration. This is because the release binary for one platform may overwrite another (for example, macOS and Linux binaries have the same name when compiled).

To see how this works, we'll launch a terminal on a macOS computer and will compile and package applications for the current macOS system, followed by Windows and Linux. You can use any project; the following example uses the package example from earlier in this chapter. Let's see how this goes:

1. First of all, we check that the application is building correctly for the current system. For our macOS host computer, this will create a package.app file, as the application we are building is called package:

```
$ ls
Icon.png main.go
$ fyne package
$ ls
Icon.png      main.go      package      package.app
```

Before moving on, we should remove any temporary files, and as we are just testing we can remove the packaged app we created as well:

```
$ rm -r package.app package
```

2. Next, we will build for Microsoft Windows. As described in the preceding table, this will require the installation of the `mingw-w64` package using Homebrew or another package manager. With this installed, we can set the environment variables, including `CC` to specify the compiler. The command will look like the following:

```
$ GOOS=windows CC=x86_64-w64-mingw64-gcc CGO_ENABLED=1
fyne package
$ ls
Icon.png    fyne.syso    main.go    package.exe
```

As you can see, this successfully built the `package.exe` file as well as a `.syso` temporary file (this is what Windows builds use to bundle metadata – it can normally be ignored).

Before packaging for Linux, we will remove these files:

```
$ rm package.exe fyne.syso
```

3. Preparing a Linux build from macOS requires more work. First, you will need to install the compiler toolchain, which is the `FiloSottile/musl-cross/musl-cross` package in Homebrew. After this, you will need to locate and install suitable X11 and OpenGL packages for Linux development (the details here will vary based on the Linux computer you are building for; detailed information can be found in *Appendix C, Cross-Compiling*). Once this is all installed, you can execute the Linux build much like the Windows command previously, but using the appropriate `CC` variable:

```
$ GOOS=linux CC=x86_64-linux-musl-gcc CGO_ENABLED=1 fyne
package
$ ls
Icon.png    main.go    package    package.tar.gz
```

And so, you can see that it is possible to build for all different platforms from a single development computer.

However, this involved a lot of package installation and configuration. To avoid this complication there is a helpful tool, `fyne-cross`, which packages the required files for easier cross-compiling.

For each of the preceding builds, we could also have specified a GOARCH variable if we wanted to target, for example, a 32-bit computer while building on our 64-bit desktop. Likewise, specifying an ARM architecture allows us to compile for Raspberry Pi computers.

> **Note that the iOS and Android targets do not use a traditional toolchain**
>
> The ability to build for mobile targets is provided by the platform-specific development packages (for example, Xcode or the Android SDK). This means you can avoid the manual compiler configuration, but will need to use the fyne package instead of a traditional go build process.

Using the fyne-cross tool

The fyne-cross tool was created to provide a simple cross-compiling approach for the Fyne toolkit. It utilizes a Docker container to package all of the build tools so that the developer does not have to install them all manually, as we did in the previous section. You can read more about the project at https://github.com/fyne-io/fyne-cross.

Using fyne-cross, you can simply specify the *platform* you would like to build for on the command line and the tool sets up the development environment and builds the package as requested. The platform parameter is like the -os parameter we used earlier.

To be able to install and use this tool, all we need is our existing Go compiler and an installation of **Docker** (an application that manages software containers). We will now step through the work involved to build our first app using fyne-cross.

In this example, we have a macOS computer building for Linux (the configuration that was complex in the previous section):

1. First, we must install Docker. The easiest way to do this is to download and run the desktop installer from their website at https://docs.docker.com/get-started/. Unfortunately, this is not supported on Linux, so you will need to install Docker Engine (usually in the package named docker) using your package manager.

2. To run `fyne-cross`, the Docker app must be running. If using Docker Desktop you should see the icon in your system tray (see the icon on the left in *Figure 9.3*). If it is not running, then just open the app using its launch icon (the icon on the right in *Figure 9.3*):

Figure 9.3 – The Docker running symbol and app icon

If running on Linux, then make sure that the service is started according to your specific distribution's documentation.

3. To install the `fyne-cross` tool, we use a version of the `go get` command, which will install it along with other Go-based applications in the `~/go/bin/` directory:

```
$ go get github.com/fyne-io/fyne-cross
```

4. Next, we issue the command to run `fyne-cross`. The basic build requires a single parameter that is the operating system we want to build for, so for Linux, we simply call the following:

```
$ fyne-cross linux
[i] Target: linux/amd64
[i] Cleaning target directories...
[√] "bin" dir cleaned: /.../Chapter09/package/fyne-cross/
bin/linux-amd64
[√] "dist" dir cleaned: /.../Chapter09/package/fyne-
cross/dist/linux-amd64
[√] "temp" dir cleaned: /.../Chapter09/package/fyne-
cross/tmp/linux-amd64
[i] Checking for go.mod: /.../Chapter09/package/go.mod
[i] go.mod not found, creating a temporary one...
Unable to find image 'fyneio/fyne-cross:base-latest'
locally
base-latest: Pulling from fyneio/fyne-cross
(downloads lots of stuff)
[√] Binary: /.../Chapter09/package/fyne-cross/bin/linux-
amd64/package
[i] Packaging app...
[√] Package: /.../Chapter09/package/fyne-cross/dist/
linux-amd64/package.tar.gz
```

5. Once this has completed (the first run will take some time as the Docker image needs to be downloaded), we should see that the expected package has been created for us:

```
$ ls fyne-cross/dist/linux-amd64
package.tar.gz
```

As you can see, the `fyne-cross` tool was able to create the application package for a system that was otherwise difficult to compile for.

Builds for any operating system and platform in the supported list (at the time of writing) include the following:

- `darwin/amd64`
- `darwin/386`
- `freebsd/amd64`
- `linux/amd64`
- `linux/386`
- `linux/arm`
- `linux/arm64`
- `windows/amd64`
- `windows/386`
- `android`
- `ios`

> **Note**
>
> iOS compilation is supported only on macOS computers. You will need to download and install Xcode from the Apple App Store. This is a restriction of Apple licenses and unfortunately cannot be worked around.

If you are able to install Docker, this is probably the easier way to build for different computers.

Summary

In this chapter, we have seen the steps involved to take an application from running from the source code, through to packaged files ready for distribution. We saw the techniques and tools available to help make applications portable and how the Fyne toolkit can offer hints of how to improve your UX.

We also explored the world of cross-compiling and how to create application packages for different operating systems. As illustrated in this chapter, it is possible to set up your development computer to build for all supported platforms; however, we saw that this can be complicated. The fyne-cross tool was introduced as a way to solve this complexity and make it trivial to package builds for the multitude of potential target systems.

In the next chapter, we will look at how to distribute these files. We will explore how you can share packaged files with beta testers and then how to prepare the packages with the certification required for app store and marketplace uploads.

10

Distribution – App Stores and Beyond

The final challenge in cross-platform development is how to deliver your app to distribute your completed project. Whether you want to distribute using platform-specific app stores, package managers, or through a simple download site, there is a little more work to do. This chapter explores the steps required to deliver applications to system app stores, mobile marketplaces, and download sites.

This chapter will cover the following topics:

- Building your application for release
- Distributing apps to desktop app stores
- Uploading apps to Google Play and iOS App Store

By the end of this chapter, you should have the knowledge to distribute your apps to any platform.

Technical requirements

This chapter has the same requirements as *Chapter 3, Window, Canvas, and Drawing*, that is, to have the **Fyne** toolkit installed and a **Go** and **C** compiler working. For more information, please refer to that chapter.

For deployment to **Android** devices, you will need to install the **Android SDK** and **NDK** (see *Appendix B, Installation of Mobile Build Tools*). To build for **iOS** devices, you will also need to install **Xcode** on your **Macintosh** computer (an Apple Mac is required for licensing reasons).

The full source code for this chapter can be found at `https://github.com/ PacktPublishing/Building-Cross-Platform-GUI-Applications-with- Fyne/tree/master/Chapter10`.

Building your application for release

As we saw in *Chapter 9, Bundling Resources and Preparing for Release*, the `fyne package` command bundles our application binary and metadata into a format that can be installed on **operating systems (OSes)** just like any native graphical application. However, in addition to differing formats for each platform, there are additional considerations when looking to distribute software, such as certification and app store upload file formats. To handle this, there is another sub-command within the build tool, `fyne release`.

In this section, we will learn how to use the `release` command to prepare an application for sharing.

Running the release command

Just like the `fyne package` command that we saw in *Chapter 9, Bundling Resources and Preparing for Release*, this new `release` command is responsible for packaging up our application with its metadata. The `release` command, however, applies changes to the application in preparation for distribution. The specific changes will differ depending on the OS, but usually include the following:

- Turning off any debug output
- Instructing applications to use *production* identities for web services
- Packaging applications into distribution-specific archives
- Applying the certification required for app stores

As we work through this chapter, we will look at options that are specific to each platform, but in this section, we can explore how to adapt applications for release builds for any OS.

Let's imagine that we have a function in our app called connectToServer() that will initiate a web connection to one of our company's services. Throughout the development process, it has been connecting to a development server, but for our distributed app we want to use the **production** (sometimes called *live*) server.

The following steps demonstrate how we can use this type of build to adapt the code appropriately for release:

1. To build this demo, we create a new main.go file that defines two different possible server authentication keys, serverKeyDevelopment and serverKeyProduction:

```
const (
    serverKeyDevelopment = "DEVELOPMENT_KEY"
    serverKeyProduction  = "PRODUCTION_KEY"
)
```

2. Next, we add a simple function that opens a dialog window showing the authentication key that will be used:

```
func connectToServer(a fyne.App, w fyne.Window) {
    key := serverKeyDevelopment
    if a.Settings().BuildType() == fyne.BuildRelease {
        key = serverKeyProduction
    }

    dialog.ShowInformation("Connect to server",
        "Using key: "+key, w)
}
```

As you can see, this function takes the current fyne.App as one parameter, so that we can query the build type using the BuildType() function. The second is the current fyne.Window parameter, which we use so that we can show the dialog for this example. Such a function would normally return a server connection, but this is just a simple demo.

3. Just like previous examples, we also need to create a basic `main` function that will run the example. In this case, we open a new window that says simply **Connecting...**, and then we launch the `connectToServer` method that will demonstrate the build type:

```
func main() {
    a := app.New()
    w := a.NewWindow("Server key demo")

    w.SetContent(widget.NewLabel("Connecting..."))
    w.Resize(fyne.NewSize(300, 160))
    connectToServer(a, w)
    w.ShowAndRun()
}
```

4. Now we simply run the application:

```
$ go run .
```

You should see that it creates a window, shown in the following screenshot:

Figure 10.1 – Running in development mode

5. Next, let's build the application for release. As we are packaging the app metadata at this point, we will need an icon file. An example, `Icon.png`, is included in the book repository, but you can add any you like, placing it alongside `main.go`. We can then prepare the release version using the `fyne` tool:

```
$ fyne release
```

6. Running the packaged application by double-clicking the application, you should see this different output:

Figure 10.2 – Running in release mode

The preceding example is a simple introduction to the `release` command and how it can adjust application behavior. As we use it throughout this chapter, you will see additional parameters required by various systems for certification and other features. We could have added information here about the version and build number (using `-appVersion` and `-appBuild`), but for most desktop releases this is optional. We will start adding it to the commands later in this chapter.

With the application now bundled in this section, we could distribute the application to *beta testers* or small communities who are happy to manage their own software installation by downloading from the web.

Sharing your app on the web

With an application now set up to use the production values where appropriate, we can start a plan for distribution. In many cases, the files created by `fyne release` will be in a different format to `fyne package` (because this command focuses on app store and marketplace distribution). To apply the release parameters to the previous package format, you can use `fyne package -release` instead.

If you want to share your applications without using the platform's official distribution, then this can be done by uploading the result of the `release` command to a website or file sharing platform. In most cases, this is a simple case of copying the file or uploading it to wherever you want to share it. However, with some systems (mainly macOS), the application is a bundle, or directory, and this may not be downloadable from a website link. In these situations, it is a good idea to compress or archive the set of files into a single file that can be downloaded.

On your mac (where macOS apps are normally created) you can open the app folder in **Finder** and right-click the package, and then select the **Compress <filename>** option from the context menu, shown as follows:

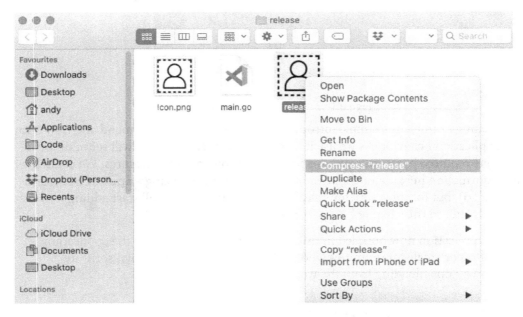

Figure 10.3 – Compressing a macOS app bundle for sharing on the web

The resulting `.zip` file created can be shared more easily from a website, and people downloading it can just double-click the file to expand your application, which can then be run as normal.

In this section, we saw how to package an application for release and how to share the resulting file so that others can install our software. However, for most applications, it will be more beneficial to distribute through the provided store or marketplace. We'll step through this process for the most common OSes in the following sections.

Distributing apps to desktop app stores

Most desktop OSes now have a central location for discovering and installing applications. Apple created the **Mac App Store**, Windows has the **Microsoft Store**, and each Linux distribution has its preferred package manager. Having an application listed in (and hosted by) a platform marketplace significantly increases the number of users you can expect and also reduces the associated hosting costs. When paired with carefully prepared metadata (as described in *Chapter 9, Bundling Resources and Preparing for Release*), a marketplace can easily become your largest distribution channel.

In this section, we step through the process for the macOS and Windows stores. We will return to Linux and BSD distribution later in the chapter as they are less mainstream and far more complicated.

Mac App Store

The Mac App Store is the desktop version of Apple's famous iOS App Store. It provides many thousands of applications available to buy and download, or gift to others. There is also curated content, which includes listings of the most popular apps in various categories, as well as staff picks and recommended software. Unfortunately, the Mac App Store cannot be browsed online as it requires the App Store software, which is pre-installed on compatible Mac computers.

To distribute an app on the App Store, you will need to have the development tools installed (see *Appendix B, Installation of Mobile Build Tools*), but you will also need to sign up to the **Apple Developer Program**. If you are not already a member, you can sign up at developer.apple.com/programs/enroll/. The development resources are free to access, but there is an annual subscription charge for access to the code signing tools, which are required to complete the release process we are about to explore.

Packaging a macOS release

To package a release for upload to the Mac App Store is like the release process we explored earlier in this chapter, but we must also provide certification details that apply **code signing** to our application.

Code signing is a complicated process to set up, so for the purpose of this description, it is assumed that you already have a distribution certificate installed. For further instructions on downloading your certificate, please see Apple's documentation at developer. apple.com/support/certificates/. You need to note the name of the certificate (use **Keychain Access** to find your developer certificates) for use in the build command later. You will also need to note which provisioning profile you are using to build. As with certificates, this should already be downloaded and installed on your computer, but you need to note the name (you can find more details on the Apple Developer portal at developer.apple.com/account/resources/certificates/).

Once you have identified the certificate name and provisioning profile, you can use the details in the following command:

```
$ fyne release -appVersion 1.0 -appBuild 1 -certificate
"CertificateName" -profile "ProfileName"
```

The resulting app package is ready to be uploaded to the **App Store Connect** website for validation.

Uploading to the Mac App Store

App Store applications are managed through the App Store Connect website (at `appstoreconnect.apple.com/`). Log in using your Apple Developer account and create a new application (if you have not already done so). This is where you add the metadata that will be displayed in the store – be sure to check the information carefully as some data cannot be changed after release.

Well-chosen descriptions and screenshots will help your application to be more easily discovered. Within this app definition, you need to start preparing a new release, with an appropriate version number and supporting information. You will probably notice that you are not yet able to select a build – to enable this, we first need to upload the compiled package.

Apple has recently created a new app named **Transporter** that is currently the easiest way to upload a new build. You can follow these steps to do so:

1. Open the Transporter app and log in with your **Apple ID**.

2. Once logged in, you will be asked to choose the application to upload – select the application package created by the `release` command in the previous section and proceed to the upload.

3. Once complete, the build will appear on the App Store Connect website (you may need to refresh the page). If you prefer command-line tools for managing the upload progress, you can use **xcrun altool**, which provides the same capability. Whichever way you upload the package, you need to select the resulting build in the app details on the website.

4. Once you have chosen this new build, you can click on the **Submit for Review** button to start the review process.

We'll now move to the next section to learn about the review process.

Mac App Store review process

As soon as an app is submitted for review, it goes through an automated set of code checks. This process verifies that the application does not contain obvious errors in metadata or code signing, and performs a code analysis to ensure you are not using APIs that are private to Apple or otherwise restricted. Assuming that these automated checks pass, then the application will be sent for final acceptance by a member of the App Store review team.

The review team checks your application for quality, reliability, compliance with the **Human Interface Guidelines (HIG**: developer.apple.com/app-store/ review/), and that it meets other criteria for inclusion on the store. This process typically takes a day or two, but can be longer for the first release of a new application. Once the process is complete, your software will be available to purchase or download on the App Store. In your first week of distribution, it may even be included in the **New and Noteworthy** section.

Now that we have understood the Mac App Store process, let's look at the Microsoft Store process.

Microsoft Store

The **Microsoft Store** is the official location for finding and installing software, apps, and games for all the current **Windows, Windows Phone**, and **Xbox** devices. It provides hosting and search facilities as well as handling payments for paid software and supports discounts and vouchers as well. You can browse the contents of the Microsoft Store online (at www.microsoft.com/store/apps) or by using the Store apps on each of the systems it supports.

To submit an application to the Microsoft Store, you will require a Microsoft account (which you may already have if you've signed in to Windows, Xbox, or **Office 365**). You will also have to start an annual subscription to access the relevant portions of the partner portal. You can log in and sign up at partner.microsoft.com/en-US/ dashboard/apps/signup.

Packaging a Windows release

Preparing a release package for the Microsoft Store is very similar to the packaging we have done before, but as a release asset, it will require version numbers applied. As with the macOS App Store, we will need to sign the software. If you have not already done so, be sure to download your certificate, noting its name. Windows apps use a combined version number, such as `1.2.3.4`, but the `fyne` tool splits this into two parts. To package the app for a specific version, pass the last number as `-appBuild` and the rest of the version as `-appVersion`, shown as follows:

```
$ fyne release -appVersion 1.2.3 -appBuild 4 -certificate
"CertificateName" -password "MyPassword"
```

The output of this command will be a `.appx` file. This file is the package type required for uploading to the store and should not be shared in other ways.

Uploading to Microsoft Store

The completed package should be uploaded to the partner portal within the **Packages** page. When preparing for upload, make sure that all of your application metadata has been added to the correct locations so that people will find your software easily. Once the package is uploaded, it will be checked for various errors that could stop it from being released. Should you encounter any warnings, you will need to remove the uploaded build from the portal, fix the issue, and upload a new package for retesting.

Microsoft Store review process

Once your package is uploaded and has passed the initial validation, it will be added to a queue to be reviewed. The Microsoft staff will review your application for correctness and suitability, and validate that it is of suitably high quality for inclusion in the store. Assuming that these checks all pass, they will publish it for distribution across the devices that you specified during the submission process.

In the next section, we will learn about Linux and BSD distributions.

Linux and BSD distributions

Linux and BSD distributions have a reputation for handling package distribution well; a desktop system will likely have a graphical package management application that provides easily searchable indexes containing thousands of packages. Despite there being hundreds of different Linux and BSD distributions, they are all based on similar binary files and file structures.

Packaging for Linux/Unix

The basis of packaging Fyne apps for Linux and BSD systems is the `.tar.gz` file created in the *Building for release* section at the beginning of this chapter. The contents of that file will be adapted during packaging for a target distribution. By adding platform-specific metadata and running the file's package command, you can create packages for your preferred distribution.

Review process for Linux packages

Once packages have been created for a system, the application developer can submit it to the package list. For each system, the method of requesting addition is different, but can normally be found in the project documentation. However, as Linux and BSD are open source systems, you may find that an existing package maintainer may be happy to do that for you.

In this section, we saw how Linux and BSD packaging can be accomplished using the available tools. This, combined with the *Mac App Store* and *Microsoft Store* sections, cover how to distribute your apps on most of the major desktop OSes. Next, we will see how to adapt this approach for uploading mobile applications to the appropriate distribution channels.

Uploading apps to Google Play and the iOS App Store

Managing releases for mobile platforms can be slightly more complicated as the apps cannot be directly tested on the devices that we use to build them. However, the Fyne release process is designed to encompass all platforms, so in this section, we will see how the same tools can be used for releasing to mobile device stores.

iOS App Store

Packaging applications for iOS is very similar to the macOS distribution process described earlier in this chapter. You will need to have **Xcode** installed and a distribution certificate and provisioning profile prepared and downloaded from the Apple Developer portal, at `developer.apple.com/account/resources/certificates/`.

Packaging an iOS release

For an iOS `release` package command, we run the following command on our Mac system, remembering to add the `-os ios` parameters:

```
$ fyne release -os ios -appVersion 1.0 -appBuild 1 -certificate
"CertificateName" -profile "ProfileName"
```

The file resulting from the preceding command will be a `.ipa` file, which is the format required for uploading an iOS app to the App Store. Once you have this file, it can be uploaded using the same process described in the *Mac App Store* section earlier in this chapter.

iOS App Store reviewing process

The review process for iOS apps is very similar to the Mac App Store. Submission of a build will start an automated check process that usually takes just minutes. After this, a manual review will be performed that can last for days, but even for reviews of complex apps in busy periods, the review will normally be completed within a week.

In the next section, we will look at the Google Play Store process.

Google Play Store

Packaging an Android app for the Google Play Store requires that we apply versioning information and our certification to the package file. Once again, the `fyne release` command will handle this, but we need to gather the relevant information. Version information can be passed the same as for previous builds, but the certification is different for Android.

Packaging an Android release

Certification information for Android apps is stored in a `.keystore` file. If you have distributed an Android app before, you will probably already have one. If this is your first time, you will need to set up your developer credentials, including a private key and certificate. More information is available in the official documentation at `developer.android.com/studio/publish/app-signing`. Be sure to keep your **keystore** safe as it will be required to release future versions of your app.

Once you have located the keystore that contains this information, you are ready to start the release process. Using the same version information as other builds, we will run the `fyne release` command with an additional `-keyStore` parameter that is followed by the path to the file. The full command will look something like the following:

```
$ fyne release -os android -appVersion 1.0 -appBuild 1
-keyStore "path/to/file.keystore"
```

Be sure to have the password and optional key alias at hand as the command will ask for additional information to complete the signing process. The result of this will be a `.apk` file much like from previous development packaging, but with signing details ready for upload. You might consider renaming the file to mark it as a release so there is no confusion later when you start with the development again.

Uploading to Play Console

To upload your application for inclusion in the Google Play Store, you will need to log in to the **Play Console** (there will be a small charge if you have not already signed up) at `play.google.com/console/`. Once logged in, you can start the process of creating a new application. Here you will upload the metadata, icons, and other marketing materials, as with the Apple iOS App Store explored earlier.

In the **Production Releases** section of the Play Console, you can start to define a new release for this app. Be sure to use the same version number as in the `release` command so that the file matches the metadata online. Once you have filled in the information, locate the **App Bundles and APKs** panel where you can drag the `.apk` file (or tap on the **Upload** button to choose it). Once the upload has completed, it will go through some automated checks to confirm compatibility. If anything is wrong at this stage, you can go back to your project to make the changes, re-build the file in release mode, and upload a replacement file.

Play store reviewing process

Once your app has been submitted to the Google Play Console, you can track its progress through the review process. There are some manual aspects to the checks that will be performed, which can lead to wait times of around a day or two. **Google** notes that in some cases, additional checks may need to be performed that can take up to 7 days or longer. This can often be the case with new user accounts or apps when they are first being published. Once approved, your app will be visible in the Play Store on all supported Android devices.

Summary

In this final chapter, we looked at how to package and distribute graphical applications using the Fyne tool. Unlike the distribution of command-line or system utilities, the process of delivering a GUI application requires additional metadata and packaging to integrate well with each OS. We saw how the basic release process can create an application package ready for distribution beyond our development and test team members.

Packaging for different platforms can be complicated, so we walked through the steps required to build native-looking graphical packages for macOS, Windows, and Linux, as well as mobile packaging for iOS and Android. Each package has its own metadata format and package structure, but this was generated automatically using the `fyne release` tool.

We also saw how to build release packages for distribution on official stores, and how these packages can be submitted to the app store or marketplace that will be preinstalled on the user's device. The Windows, macOS, iOS and Android stores provide an opportunity for applications to earn revenue following release, and the Linux software listings will help to increase the visibility of our software package.

Having learned the various aspects of developing graphical apps using Fyne, we have made it all the way to a complete and published application. Hopefully, this guide has been helpful and has enabled you to create the app that you had been aiming to build and to get it working across all your devices over a range of OSes.

If you would like to distribute your application to the community, please consider sharing it on the Fyne app listing by going to `developer.fyne.io/submit`.

Appendix A: Developer Tool Installation

In preparation for running the code examples in this book, you will need to have both the **Go** compiler and a **C** compiler installed (to support **CGo**). If you don't have either of these set up, this appendix will guide you through the installation.

Installing on Apple macOS

Many developer tools (including **Git**) are installed as part of the **Xcode** package. If you haven't already installed Xcode for other development work, you can download it for free from the **Mac App Store** at apps.apple.com/app/xcode/id497799835.

Once installed, you should also set up the command-line tools. To do this, go to the Xcode menu | **Preferences** | **Downloads** | **Install Command Line Tools**. If you're unsure about whether you've installed these already, then open the **Terminal** application and execute xcode-select. If the tools are already installed, this will execute normally. If not, you'll be prompted to run the installation, as shown in the following screenshot:

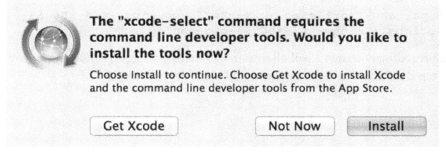

Figure 11.1 – The installation dialog window when developer tools are not installed

In addition to these tools, you'll need to install Go. You can get the download package from `golang.org/dl`. From there, tap on the featured download link for Apple macOS, and then run the installer package that downloads. You may need to close any open **Terminal** windows to update your environment variables.

Installing on Microsoft Windows

Configuring a development environment for Windows can be complicated as there aren't many tools installed by default. Due to this, there are many options for setting up, including using external tools and packages (such as **MSYS2**, **MinGW**, and **Windows Subsystem for Linux**). However, exploring these is outside the scope of this book.

The following steps show you how to get up and running using MSYS2, which provides a dedicated command-line application that will be set up for Fyne development. Let's get started:

1. You will need to download the installer from `www.msys2.org`. You should choose either the 32-bit (`i686`) or 64-bit (`x86_64`) version, depending on your computer architecture.

2. Once downloaded, run the installer, which will download the basic packages on your computer, including the package manager (`pacman`).

3. Once the installation is complete, you'll be given the opportunity to launch the **MSYS Command Prompt – please do not accept this**, as this is not the version of the app we want to run.

4. Once completed, open the directory you chose to install the application into and run the `mingw64.exe` application instead. This is the command line that comes pre-configured with knowledge of Windows compilation. We can now use the package manager to install Go and Git, as well as the C compiler toolchain and `pkg-config` (which is used by CGo to find packages):

```
$ pacman -S git mingw-w64-x86_64-go mingw-w64-x86_64-
toolchain mingw-w64-x86_64-pkg-config
```

The preceding command will offer to install many packages, which is want we want. Just tap the *Return* or *Enter* key to install these dependencies:

Figure 11.2 – Running the installed Mingw64 terminal to install extra packages

Note that the preceding terminal prompt denotes **MINGW64**. If you see **MSYS2** or another prompt, then you have opened the wrong terminal application.

Once these packages have been installed, you will have a full development environment. The default Go home path will be `C:/Users/<username>/go`, though you should consider adding `C:/Users/<username>/go/bin` to your `%PATH%` environment variable.

Installing on Linux

Setting up the prerequisite software on Linux should only require installing the correct packages for your distribution. The `git` package will provide the source control tools, and the Go language should be in the `go` or `golang` package. Additionally, the CGo requirement means that the `gcc` package will need to be present as well. Installing these packages will provide the necessary commands to run the examples in this book. You may need to add `~/go/bin` to your `PATH` environment variable to be able to run tools that Go installs later.

There are various different package managers for Linux, and each have slightly different naming conventions for packages, as well as different commands. The following commands are examples of how to install the required packages for each of the most popular distributions. Some platforms require additional library headers or dependencies to be installed, which are included where required:

- **Arch Linux**: `sudo pacman -S go gcc xorg-server-devel`

- **Fedora**: `sudo dnf install golang gcc libXcursor-devel libXrandr-devel mesa-libGL-devel libXi-devel libXinerama-devel libXxf86vm-devel`

- **Solus**: `sudo eopkg it -c system.devel golang mesalib-devel libxrandr-devel libxcursor-devel libxi-devel libxinerama-devel`

- **Ubuntu / Debian**: `sudo apt-get install golang gcc libgl1-mesa-dev xorg-dev`

- **Void Linux**: `sudo xbps-install -S go base-devel xorg-server-devel libXrandr-devel libXcursor-devel libXinerama-devel`

After using the aforementioned commands, you will have a full Fyne development environment ready on your computer.

Appendix B: Installing Mobile Build Tools

Due to the way that mobile applications are compiled, they require additional tools and packages for testing or installation on devices. In this appendix, we'll learn how to set up the additional components for **iOS** and **Android** development.

Preparing for Android

To develop apps for Android, we will require additional development tools. Development follows the same process as we have seen in the earlier chapters of this book, and the same Fyne APIs are available for your application – it's just the build/package phase that changes. Here are the necessary steps that we need to follow:

1. Firstly, you will need to install the Android SDK. The easiest way to do this is by installing **Android Studio**, which is available at `developer.android.com/studio`. Tap the **download** button on that website and follow the installation instructions for your type of computer.

2. Once the installation has completed, you will also need to install the **Native Development Kit (NDK)**, which is managed through the **SDK Manager**. This can be accessed through Android Studio by accessing the **Tools** menu and choosing **SDK Manager**. If you are not using Android Studio, then the same functionality can be seen by running the `sdkmanager` application.

3. Once this has loaded, choose **SDK Tools** to see the full list of tooling available. Here, you will need to check the **NDK** option (sometimes called **side by side**) and **Android SDK Tools**, if they are not selected. Tap the **Apply** button and the packages will be installed, as follows:

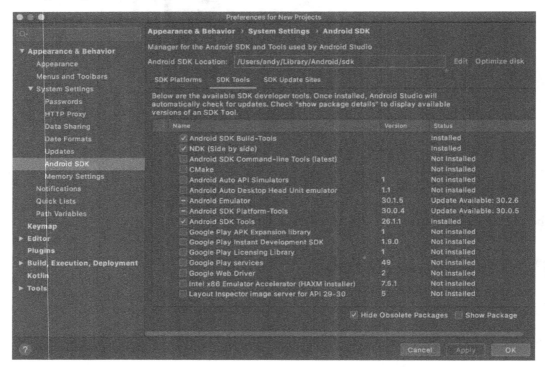

Figure 12.1 – The Android SDK manager (showing that the necessary SDK tools and NDK are installed)

4. Once this has been set up, you should be able to build any Fyne app for your Android phone or tablet device using a command such as the following:

```
$ fyne package -os android -appID com.example.myapp
```

A successful command will create an `.apk` file that can be installed using the Android tools available to you, the `adb install` command, or the `fyne install` command.

It is also possible to use `fyne-cross` for Android packaging, as shown in *Appendix C: Cross Compiling*.

Preparing for iOS and iPadOS

Building apps for **Apple** mobile devices is a little different from compiling your applications for macOS (desktop computers). Firstly, this must be done on an Apple Macintosh computer (**iMac**, **MacBook Pro**, and more) due to licensing restrictions.

Secondly, you must have Xcode tools installed (this is somewhat optional for desktop app creation). Lastly, if you want to test on physical devices or distribute to the store, you will need to be signed up to the **Apple developer program**, which carries an annual cost.

The Xcode installation is described in the *Apple macOS* section of *Appendix A: Developer Tool Installation*. If you have previously done iOS development, this will already be set up.

Next, you will need to have an Apple developer account. If you have not done so already, you can sign up at developer.apple.com/programs/enroll/. There is an annual fee, and if you publish apps but fail to renew your membership, they will be removed from the store. Once signed up, you should follow the documentation to set up a **developer certificate** and add your devices to the account device list so that you can use them for testing. Once added, you will need to create a **provisioning profile** – a single wildcard profile is sufficient for general development.

Once you've done this, you should be able to build any Fyne app for your Apple phone or tablet device using a command such as the following:

```
$ fyne package -os ios -appID com.example.myapp
```

Successfully executing the preceding command will create a .ipa file that can be installed on your devices for testing. This process is covered in more detail in *Chapter 9, Bundling Resources and Preparing for Release*.

Appendix C: Cross-Compiling

When building applications that need access to native APIs and graphics hardware, we can use **CGo**. Although not much harder to build for regular development, this does make cross compiling much more complicated. For every target platform you want to build for, there must be a **C compiler** that knows how to create native binary files. This appendix outlines the steps required to set up cross compilation targets for each combination referenced earlier in this book.

> **Important note**
>
> Please note that cross compiling is not required for day-to-day development. For most development, you won't require the cross-compiler setup. The Go compiler and standard tools discussed in *Appendix A: Developer Tool Installation*, are all that you require to develop for standard computers. This appendix is all about installing additional tools for creating compiled applications for a different operating system or architecture than your current computer.

We will cover two possible approaches for compiling in this appendix:

- Manually installing the cross-compiler toolchain
- Using `fyne-cross` to handle the compilation automatically

We will start with the manual process since it is useful to understand the compiling complexity before seeing how the process can be automated.

Manually installing compilers

Installing compilers and toolchains is complex, but this appendix will attempt to guide you through the main steps. This approach is sometimes preferred by developers who want to manage every single detail of their computer. It may also be easier if your computer has previously been used for **C** development and creating native apps for multiple platforms.

Preparing to build for a different target varies, depending on the system you want to compile for. We'll start by looking at **macOS**, before exploring **Windows** and **Linux** afterward. It is not necessary to follow these steps to build mobile apps since we installed these tools in *Appendix B: Installing Mobile Build Tools*.

Cross compiling for macOS

When cross compiling for macOS, it's necessary to install the **Software Development Kit** (**SDK**) from **Apple**, as well as a suitable compiler. The instructions for **Windows** (using **MSYS2**, as described in *Appendix A, Developer Tool Installation*) and Linux are almost identical; the main thing we need to do is install the macOS SDK.

One thing you need to be aware of is that Apple have removed support for building 32-bit binaries. If you wish to support older devices that are not 64-bit, you will need to install older Xcode (9.4.1 or lower) and Go (1.13 or lower) versions.

The easiest way to install the macOS SDK is by using the `osxcross` project. The following steps show you how to download and install the SDK and necessary tools for building a Fyne app for macOS, without using a Macintosh computer. Here, we're using Linux, but the process is the same for Windows developers using the MSYS2 command-line tools:

1. We'll be using the `clang` compiler in place of `gcc` since it's more portable by design. For this process to work, you'll need to install `clang`, `cmake`, and `libxml2-dev` using your package manager:

    * On Linux, use `apt-get install clang cmake libxml2-dev` (or the appropriate `pacman` or `dnf` commands, depending on your distribution)

    * On Windows, use `pacman -S mingw-w64-x86_64-clang mingw-w64-x86_64-cmake mingw-w64-x86_64-libxml2`

2. Next, we need to download the macOS SDK, which is bundled with Xcode. If you don't already have an Apple developer account, you'll need to sign up and agree to their terms and conditions. Using this account, log into the download site at `developer.apple.com/download/more/?name=Xcode%2010.2` and download `XCode.dmg` (**10.2** is recommended for **osxcross** targeting 64-bit distributions, though you can download 9.4 if you want to support 32-bit computers as well).

3. Then, we must install the **osxcross** tool. We'll start by downloading it with the following `git` command:

```
$ git clone https://github.com/tpoechtrager/osxcross.git
```

4. Once downloaded, move into the new directory. Using the package tool in this repository, we must extract the macOS SDK from the downloaded `Xcode.dmg` file:

```
$ ./tools/gen_sdk_package_darling_dmg.sh <path to Xcode.
dmg>
```

The resulting `MacOSX10.11.sdk.tar.xz` file should be copied into the `tarballs/` directory.

5. Lastly, we must build the **osxcross** compiler extension by executing the provided build script:

```
$ ./build.sh
```

6. Once we've done this, a new directory named `target/bin/` will appear, which you should add to your `PATH` environment variable. The compiler can now be used in **CGo** builds if we set the `CC=o64-clang` environment variable; for example:

```
$ CC=o64-clang GOOS=darwin CGO_ENABLED=1 go build .
```

More details about this process and how to adapt it for other platforms are available on the **osxcross** project website at `github.com/tpoechtrager/osxcross`.

Cross compiling for Windows

Building for Windows from another platform requires that we install the `mingw` toolchain (this is similar to what we installed on Windows to support CGo). This should be available in your package manager with a name similar to `mingw-w64-clang` or `w64-mingw`, but if not, you can install it directly using the instructions at `github.com/tpoechtrager/wclang`.

Installing Windows tools on macOS

To install packages on macOS, it's recommended that you use the **Homebrew package manager**. If you have developed on macOS before, then you probably have this installed already, but if not, you can download it from their website, brew.sh. Once Homebrew is set up, the compiler package can be installed using the following command:

```
$ brew install mingw-w64
```

Once installed, the compiler can be used by setting CC=x86_64-w64-mingw64-gcc, as follows:

```
$ CC= x86_64-w64-mingw64-gcc GOOS=windows CGO_ENABLED=1 go
build .
```

In the next section, we'll learn how to install Windows tools on Linux.

Installing Windows tools on Linux

Installing on Linux should just require finding the correct package in your distribution's listing. For example, for **Debian** or **Ubuntu**, you would execute the following command:

```
$ sudo apt-get install gcc-mingw-w64
```

Once installed, the compiler can be used with CGo by setting CC=x86_64-w64-mingw64-gcc, as shown in the following command:

```
$ CC= x86_64-w64-mingw64-gcc GOOS=windows CGO_ENABLED=1 go
build .
```

Lastly, we will look at how to compile for Linux computers using a manual toolchain installation.

Cross compiling for Linux

To cross compile for Linux, we need a **GCC** or compatible compiler that can build Linux binary files. On macOS, the easiest platform to use is musl-cross (musl has many other advantages, all of which you can read more about at www.etalabs.net/compare_libcs.html). On Windows, the linux-gcc package is more than suitable. Let's work through the steps for each of these.

Installing a Linux compiler on macOS

To install the dependencies so that we can cross compile for Linux, we'll use the Homebrew package manager again – see the *Installing Windows tools on macOS* section or the brew.sh website for installation instructions.

Using Homebrew, we can install the appropriate packages by opening a **Terminal** and executing the following commands (the HOMEBREW_BUILD_FROM_SOURCE variable works around an issue with musl-cross, depending on potentially old versions of libraries):

```
$ export HOMEBREW_BUILD_FROM_SOURCE=1
$ brew install FiloSottile/musl-cross/musl-cross
```

Once the installation is complete (this may take some time as it's building a complete compiler toolchain from source), you should be able to build for Linux. To do so, you'll need to set the CC=x86_64-linux-musl-gcc environment variable, as follows:

```
$ CC=x86_64-linux-musl-gcc GOOS=linux CGO_ENABLED=1 go build
```

The process is similar for Windows, as we'll see in the next section.

Installing a Linux compiler on Windows

Using MSYS2, just as we did in the *Cross compiling for macOS* section, we can install the gcc package to provide cross compilation for Linux:

```
$ pacman -S gcc
```

Once the installation has completed, we can tell our Go compiler to use gcc by setting the CC=gcc Environment variable. Compilation should now succeed if you follow the instructions in your current example, as follows:

```
$ CC=gcc GOOS=linux CGO_ENABLED=1 go build
```

At this point, you may see additional errors due to missing headers. To fix this, you'll need to search for, and install, the required libraries. This typically occurs because the compiler does not have built-in knowledge of how Linux desktop graphics work.

If, for example, your error states that X11 headers couldn't be found, then you would use pacman -Ss x11 to search for the right package to install. In this instance, the desired package is mingw-w64-libxcb (a Windows version of the X11 libraries), which can be installed like so:

```
$ pacman -S mingw-w64-libxcb
```

If you can't find an appropriate package, you could try the Windows subsystem for Linux. More information is available at docs.microsoft.com/en-us/windows/wsl (this brings a full Linux distribution to your Windows desktop).

Using fyne-cross

To make it easier to manage the various development environments and toolchains required for cross-compiling a project, we can use the fyne-cross command. This cross-compiling utility uses **Docker** containers to download and run the development tools of different operating systems, right from your favorite desktop computer.

Installing fyne-cross

Before you can install fyne-cross, you will need a recent version of **Go** (at least **1.13**) and must have **Docker** installed. On macOS and Windows, there is a **Docker Desktop** application that can help you manage your Docker setup, while on Linux, most distributions have a docker package readily available. More information about working with Docker is available in *Chapter 9, Bundling Resources and Preparing for Release.*

Once Docker has been installed, you can use the standard Go tools to install fyne-cross. The following command will be sufficient:

```
$ go get github.com/fyne-io/fyne-cross
```

The fyne-cross binary will be installed in $GOPATH/bin. If the commands provided in the following sections do not work, check that the binary is in your global $PATH environment.

Using fyne-cross

Once installed, the fyne-cross command can be used to build apps for any target operating system and architecture. It has one required parameter: the target system. To build a Windows executable from a Linux computer, simply call the following command:

```
$ fyne-cross windows
```

There are many options available (see `fyne-cross help` for more information). The most common two arguments are `-arch`, which allows you to specify a different architecture, and `-release`, which invokes the Fyne release pipeline to build apps that are ready to upload to marketplaces (you can find out more in *Chapter 10, Distribution – App Stores and Beyond*. To create an application that will work on smaller Linux computers, such as **Raspberry Pi**, we can invoke the following command:

```
$ fyne-cross linux -arch arm64
```

To build an Android app for release, you can use the following command:

```
$ fyne-cross android -release
```

The preceding steps should enable you to build and distribute applications for any platform with minimal investment in hardware.

Please note that if you're running `fyne-cross` to build apps for iOS, you must run it on a macOS host computer. This is a requirement of Apple's licensing.

Packt.com

Subscribe to our online digital library for full access to over 7,000 books and videos, as well as industry leading tools to help you plan your personal development and advance your career. For more information, please visit our website.

Why subscribe?

- Spend less time learning and more time coding with practical eBooks and videos from over 4,000 industry professionals

- Improve your learning with Skill Plans built especially for you

- Get a free eBook or video every month

- Fully searchable for easy access to vital information

- Copy and paste, print, and bookmark content

Did you know that Packt offers eBook versions of every book published, with PDF and ePub files available? You can upgrade to the eBook version at packt.com and, as a print book customer, you are entitled to a discount on the eBook copy. Get in touch with us at customercare@packtpub.com for more details.

At www.packt.com, you can also read a collection of free technical articles, sign up for a range of free newsletters, and receive exclusive discounts and offers on Packt books and eBooks.

Other Books You May Enjoy

If you enjoyed this book, you may be interested in these other books by Packt:

Mastering Go – Second Edition

Mihalis Tsoukalos

ISBN: 978-1-83855-933-5

- Clear guidance on using Go for production systems
- Detailed explanations of how Go internals work, the design choices behind the language, and how to optimize your Go code
- A full guide to all Go data types, composite types, and data structures
- Master packages, reflection, and interfaces for effective Go programming
- Build high-performance systems networking code, including server and client-side applications
- Interface with other systems using WebAssembly, JSON, and gRPC
- Write reliable, high-performance concurrent code
- Build machine learning systems in Go, from simple statistical regression to complex neural network

Hands-On High Performance with Go

Bob Strecansky

ISBN: 978-1-78980-578-9

- Organize and manipulate data effectively with clusters and job queues
- Explore commonly applied Go data structures and algorithms
- Write anonymous functions in Go to build reusable apps
- Profile and trace Go apps to reduce bottlenecks and improve efficiency
- Deploy, monitor, and iterate Go programs with a focus on performance
- Dive into memory management and CPU and GPU parallelism in Go

Packt is searching for authors like you

If you're interested in becoming an author for Packt, please visit authors.packtpub.com and apply today. We have worked with thousands of developers and tech professionals, just like you, to help them share their insight with the global tech community. You can make a general application, apply for a specific hot topic that we are recruiting an author for, or submit your own idea.

Leave a review - let other readers know what you think

Please share your thoughts on this book with others by leaving a review on the site that you bought it from. If you purchased the book from Amazon, please leave us an honest review on this book's Amazon page. This is vital so that other potential readers can see and use your unbiased opinion to make purchasing decisions, we can understand what our customers think about our products, and our authors can see your feedback on the title that they have worked with Packt to create. It will only take a few minutes of your time, but is valuable to other potential customers, our authors, and Packt. Thank you!

Index

R

race conditions 25, 30
radial gradient 64
RadioGroup widget 122
raster output 61
Red, Green, and Blue (RGB) 137
release
  application, building for 258
release command
  running 258-261

S

scalable drawing primitives 57, 58
scalable vector graphics (SVG) 57
Scroll container 135
Search Engine Optimization (SEO) 244
SelectEntry widget 123
Select widget 122, 123
semantic API 37
separation of concerns 228
Set(float64) error 167
Slider widget 123
smartphones 10
snake game
  implementing 66
  keys, used for controlling
    direction of snake 70-72
  movement, animating 72-74
  snake, drawing on screen 66-68
  timer, adding to move snake 68-70
Split container 135, 136
standard layouts
  about 81
  BorderLayout 86, 87
  BoxLayout 83

CenterLayout 82
FormLayout 84
GridLayout 84, 85
GridWrapLayout 85, 86
MaxLayout 81
PaddedLayout 82
standard library 29
standard theme
  customizations, providing to 208-210
standard widgets
  data, using with 169-171
strings
  types, formatting into 171, 172
string type
  values, parsing from 172, 173
structs
  mapping, to data binding 179
structure
  adding, with container widgets 133
syscall 31

T

Table widget
  about 129
  callbacks 130
  selection 130
task list application
  content, editing 158, 159
  data, defining 149-151
  details, filling 154-156
  GUI, designing 147-149
  implementing 146
  tasks, creating 161
  tasks, marking as complete 160, 161
  tasks, selecting 152, 153
template item 128

X

www.ingramcontent.com/pod-product-compliance
Lightning Source LLC
Chambersburg PA
CBHW080929060326
40690CB00042B/3227